IN PRAISE OF *Defe*

"Patrick Curry's attack on the co...
Tolkien is accurate and enjoyably ruthless. But the heart of his
book is its discussion of content: what *The Lord of the Rings* means
to the readers who, ignoring the critics, have for fifty years given
it a popular status as solid as that of *Don Quixote*. Though modernism defines fantasy as irrelevance, readers know better. Curry
shows us that they find in Tolkien a concentrated view of the
great ills and perils of our age—loss of community with the natural world, pursuit of happiness through material gain and technological empowerment, reduction of religious feeling to funda-
mentalist beliefs—and a morally subtle narrative which offers the
possibility of re-sacralization of the natural and human world.
What Tolkien offers is 'hope without guarantees,' and Curry
shows us how he does it. A most valuable and timely book."

—**Ursula K. Le Guin**

"Middle-earth finds more admirers all the time, but those who
want to know what they are admiring, and why, can do no better than to read Patrick Curry's reissued *Defending Middle-earth*.
Curry sees deeply into the spiritual heart of Tolkien's world, and
explains it in clear and unaffected language. The shade of Tolkien
would nod approval."

— **Tom Shippey, author of *Tolkien: Author of the Century***

"Wonder, reverie, enchantment, mystery, the evocative spell of
words and the magic power of places: do such experiences pertain
only to the otherworldly sphere of fantasy? Or are they pragmatic
and necessary ingredients for a life lived well in *this* world, elements integral to any truly democratic and genuinely sustainable
human culture? Such is the possibility that reverberates through
this audacious little gem of a book: a luminous study of *The Lord
of the Rings* and its growing relevance for our era. Patrick Curry's

genius is to show how creatively Tolkien's saga responds to the myriad problems threatening the health of our planet (and undermining our own humanity) at this precarious moment in history. One emerges from this book with a new respect for the keen ecological intelligence that infuses *The Lord of the Rings*, and for the transformative power of Tolkien's mythic vision. Today, as our last great forests are laid waste, as our fellow species tumble into oblivion while the earth itself shivers into a fever, we can only be grateful that such a vision, with its unique capacity for re-enchantment, is quietly spreading its influence throughout the world." — **David Abram, author of** *The Spell of the Sensuous*

# DEFENDING
# MIDDLE-EARTH

# DEFENDING MIDDLE-EARTH

## TOLKIEN:
### *Myth and Modernity*

PATRICK CURRY

HOUGHTON MIFFLIN COMPANY

BOSTON · NEW YORK

2004

First Houghton Mifflin Books edition 2004

Copyright © 1997, 2004 by Patrick Curry
ALL RIGHTS RESERVED

For information about permission to reproduce selections
from this book, write to Permissions, Houghton Mifflin Company,
215 Park Avenue South, New York, New York 10003.

Visit our Web site: www.houghtonmifflinbooks.com

*Library of Congress Cataloging-in-Publication data is available.*
ISBN 0-618-47885-x

Previously published in hardcover by St. Martin's Press

Printed in the United States of America

Book design by Victoria Hartman

MP 10 9 8 7 6 5 4 3 2 1

# Contents

## Preface

LIKE *The Lord of the Rings*, if in no other way, this tale too grew in the telling. It began life as what was intended to be a short paper for the 1992 Centenary Conference in Oxford. But I found I couldn't stop, and before long it had grown to nearly 30,000 words. Eventually, without really setting out to do so, I spent a good deal of the last five years researching and writing it. Part of the problem was that Tolkien's work impinges on so many complex and profound concerns, each with its own centrifugal tendency, that trying to exert some editorial control has been (I imagine) like riding a wild horse.

For whom is the book intended? According to Tolkien, one face of fairy-stories is 'the Mirror of scorn and pity towards Man.' Scorn and pity is certainly what I feel for one section of humanity I encountered while working on it, namely all those editors, critics and authors who look down their nose at Tolkien, usually without ever having really read or thought about him. They exhibited disturbing signs of a group 'mind' based entirely on snobbery and prejudice. I would therefore be astounded, albeit pleasantly so, if my book persuades any of them otherwise; but in any case, it is not really meant for them. May it rather provide a way in for those who come to Tolkien's books with an open mind, send former readers back to them again, and help those who already love them to appreciate them still more. (And if any of the latter have felt ashamed to admit it, may it help you understand why you need do so no longer!)

I may as well also mention that alongside this book another text has sprung up willy-nilly. Too long for an article, too short for a book—although still growing—it takes the *via negativa* of tackling

Tolkien's Marxist, modernist, feminist and psychoanalytic critics directly. This enables some things to be discussed that were not appropriate to include here. I am thinking of calling it something like, 'The Discursive Dynamics of Tolkien's Fantasy: A Post-Axiological Critique,' and seeing what happens. (Academic editors: please confound my expectations, and form an orderly queue.)

As a result of my experience with the 'experts' to date, I am all the more grateful to Christopher Moore, of Floris Books, for his thorough, sensitive and patient editorial guidance; and to Suzanna Curry, for her support, wisdom and forbearance. They have made it a much better book than it would have otherwise been.

It is also a pleasure to thank these others who have helped. Michael Winship, Clay Ramsay, Jesper Siberg, Virginia Luling, Simon Schaffer, Elaine Jordan, John Seed and Angus Clarke read and commented on early drafts. Stratford Caldecott, Tony Linsell, Kathleen Herbert, Charles Coulombe, Raphael Samuel, Dwayne Thorpe, Ray Keenoy and my mother, Noreen Curry, provided valuable moral support and encouragement in discussions or correspondence. Tom Shippey, Ursula Le Guin and Brian Attebery also made encouraging noises early on in correspondence, which I greatly appreciated. Charles Noad kindly read the whole penultimate draft and saved me from many small and some not-so-small errors; so did Nicola Bown, who made some excellent suggestions. (None of these people, obviously, have any responsibility for what I have finally said.)

I would also like to thank Warwick and Linda of the Q.E.L. Cafe, in Notting Hill Gate, for their hospitality; and Skrét, Filip and Karel of the Brno Tolkien Society, whom I met in Oxford in 1992, for reminding me what it's all about. (May you too be lucky enough to be reminded, when you have almost forgotten.)

I am also grateful to HarperCollins UK for permission to quote from the published works of J.R.R. Tolkien.

Finally, this book is dedicated to the memories of three people, very different but all with a place in my heart, who died while I was still writing it: my father, Peter D. Curry (1912–1996), the writer P. L. Travers (1899–1996), and the historian Raphael Samuel (1934–1996). R.I.P.

# INTRODUCTION:
# RADICAL NOSTALGIA

---

*People who like this sort of thing will
find this the sort of thing they like.*

IT COULD BE a literary fairy-story. A reclusive Oxford don, best-known for his scholarly work on Anglo-Saxon, unexpectedly produces a popular children's story. Seventeen years later, he follows this up with a very long story, published in three volumes. Set entirely in an imaginary world, it centres on a quest involving a magic ring and some members of a three-and-a-half-foot-tall rustic race called 'hobbits.' His book is variously described as romantic epic or juvenile fantasy; but whatever it is, it is certainly not a modern novel, and the critics are divided between bafflement and visceral dislike. The general opinion in the academic and critical neighbourhood is that, rather like one of his characters, its author, 'who had always been rather cracked, had at last gone quite mad . . .' Yet just over ten years later, his books become runaway best-sellers; and after forty years, they count among the most widely read in the global history of publishing.

The author is, of course, John Ronald Reuel Tolkien (1892–1973). Born in South Africa to English parents, he moved back to England, just outside Birmingham, at the age of three-and-a-half.

He developed a childhood passion for languages into a lifelong academic career, interrupted by service in the war of 1914–18. He became Professor first of Anglo-Saxon, then of English Language and Literature, at Oxford, where he remained for the rest of his life. Despite co-editing a respected edition of *Sir Gawain and the Green Knight*, and writing a paper on *Beowulf* acclaimed for its brilliance, it was an unremarkable life by many standards . . . except for those books.

Reliable figures are hard to come by, but as far as I can tell, total worldwide sales of Tolkien's books are as follows. *The Lord of the Rings* (1954–55), at about 50 million copies, is probably the biggest-selling single work of fiction this century. *The Hobbit* (1937) is not far behind, at between 35 and 40 million copies. And one could add the considerable sales, now perhaps over 2 million, of his dark and difficult posthumously published epic *The Silmarillion* (1977). The grand total is thus well on its way to 100 million. Tolkien's books have been translated into more than thirty languages, including Japanese, Catalan, Estonian, Greek, Hebrew, Finnish, Indonesian and Vietnamese. (This last, unofficial translation appeared in 1967, whereupon the South Vietnamese II Corps was rather perceptively fêted by tribesmen with shields bearing the Eye of Sauron.)

Furthermore, this is no flash-in-the-pan phenomenon, riding on the heels of the 1960s; Tolkien has outlived the counter-culture in which he first flourished. No longer fashionable, he nonetheless still sells steadily. That was undoubtedly the main reason for the purchase in 1990 of his publisher, Unwin Hyman (originally George Allen and Unwin), by HarperCollins.

Every other index points to the same conclusion. In England, for example, since figures began to be kept in 1991, Tolkien's books have been taken out of public libraries around 200,000 times a year; he is one of only four 'classic authors' whose annual lending totals have exceeded 300,000 (well ahead of Austen, Dickens and Shakespeare). *The Hobbit* spent fifteen years as the biggest-selling American paperback, and *The Lord of the Rings* is still the most valuable first edition published in the second half of the twentieth century. Indeed, the latter — laboriously typed out on a bed in suburban war-time Oxford, and expected by its first publisher to lose money — is now universally acknowledged as largely responsible for the subsequent money-spinning genre of 'fantasy literature.' Then there are the extra-literary phenomena. In the 1960s and 70s, buttons and graffiti proclaiming 'Frodo Lives!' sprouted (in Quebec, it was 'Middle-earth Libre'). The title of *The Silmarillion* provided the name of an early heavy-metal band, while on the more establishment side, 'hobbit' is now entrenched in the *Oxford English Dictionary*, and a thousand 'Lothlóriens' and 'Rivendells' can be found on house-signs in suburban lanes. There is now even an area of submarine features off the southwest coast of Ireland named after Tolkien characters: hence, 'Gollum's Channel,' and so on.

In other words, we are talking about a massively popular and successful publishing phenomenon; all the more so when one of the books in question is half-a-million words long, and neither involves any big money or sex, explicit or otherwise — two ingredients now normally considered essential for bestsellers — let alone cannibalism, serial murder, sado-masochism or lawyers. (And how many such books will still be in print half a century after publication? The fate of Jackie Collins beckons.)

## The Story

This book will undoubtedly make more sense if you have already read *The Lord of the Rings*; but if you have not, or need reminding, here is a very brief synopsis. It takes place in the Third Age of Middle-earth — our Earth, but in an imaginary period a very long time ago. Frodo Baggins of the Shire, where the hobbits live, inherits a magic ring from his uncle Bilbo, who had acquired it from a fallen hobbit, Gollum, in the course of adventures recounted in

*The Hobbit.* Gandalf the Grey, a wizard, realizes that it is the One Ring, eagerly sought by its maker Sauron, the ruler of Mordor and the greatest power in Middle-earth. With the Ring, Sauron would be invincible. The only hope is to try to smuggle the Ring into Mordor and cast it into the furnace of Mount Doom where it was forged; for it cannot be destroyed in any other way, and anyone who tries to use it against Sauron would simply become another Dark Lord.

Frodo and his devoted companion Sam therefore begin the quest to return the Ring to its source. Initially, they are accompanied by the Company, including and representing the 'free peoples' of Men, Elves and Dwarves, as well as Gandalf and two other hobbits, Pippin and Merry. But the Company is soon dispersed, and from then on (most of the book), the reader follows two parallel stories: the adventures of its remaining members in the War of the Ring, as they struggle to keep Sauron occupied and distracted, and the agonizing journey of Frodo and Sam, accompanied by the treacherous Gollum.

Although Gandalf has always been its chief strategist, the war against Sauron is increasingly led by Aragorn, the hitherto unknown heir to the thrones of Arnor (now vanished) and Gondor (still the chief kingdom of resistance among Men). In its course, followed principally through the fortunes of Merry and Pippin, we meet some extraordinary places and people, both human and otherwise — including Lothlórien, the last remaining stronghold of pure Elvish 'magic,' where the powerful elven lady Galadriel lives; the fierce feudal Riders of Rohan; the Ents, sentient, talking and moving trees; Shelob, a malevolent spider-being; the nine Ringwraiths, Sauron's lieutenants; and Saruman, a corrupted wizard.

When Frodo does arrive, he is mastered at the last moment by the Ring, and claims it; but Gollum bites it off his finger, loses his balance, and falls into the Crack of Doom holding it. The works of Sauron come to a cataclysmic end, and Frodo and Sam are just saved from the wreck. Eventually, after Aragorn's coronation and wedding, and together with Pippin and Merry, they return to the Shire to find their struggles not yet over. But order is finally restored, and after a few years Frodo (who never really recovers from his ordeals) is allowed to pass over the Sea to within sight of Elvenhome, together with some of the last and greatest Elves and Gandalf. Sam remains in the Shire with his wife and family.

*The Lord of the Rings* is not really a trilogy, that being merely the publisher's device for breaking it up into manageable-sized volumes; it is written in six 'books,' largely following the two parallel stories. Middle-earth's languages (both written and spoken), the histories of its various peoples, calendrical systems, and some family trees are discussed in detailed appendices — all too briefly for those readers who have fallen in love with the book en route. (Those who haven't won't have gotten that far.)

## *Readers* vs. *Critics*

The first and chief riddle I want to try to unravel is therefore this: how could such a remarkably unlikely book, written by someone so removed from (and indeed hostile to) mainstream cultural and intellectual life, achieve such a huge and lasting popular success? Or, to put it another way, what are millions of readers from all over the world getting out of reading these books?

Meanwhile, the critical incomprehension continues. Among professors of English literature and readers in cultural studies, sociologists of popular culture, literary critics, and editors both journalistic and commissioning — in short, all the class of professional literary explainers — Tolkien and his readers are a no-go zone. There are a very few honourable and excellent exceptions (which, incidentally, my own work is intended not to replace but to complement). They have, however, been largely ignored within the literary community, whose silence on Tolkien — even among those whose chosen subject is fairy-tales or fantasy — is broken only by an occasional snort of derision which seems to pass for analysis.

The pattern was set by an extended sneer about Tolkien's 'juvenile trash' in 1956 by Edmund Wilson, the champion of modernism; pompously obsessed, as a contemporary put it, 'with being the Adult in the room,' Wilson is a good example of what Ursula K. Le Guin called 'a deep puritanical distrust of fantasy.' He was joined by others, notably Philip Toynbee, who in 1961 celebrated the fact that Tolkien's 'childish' books 'have passed into a merciful oblivion.' Rarely has a death been so exaggerated. But Tolkien is still routinely accused of being variously 'paternalistic, reactionary, anti-intellectual, racist, fascistic and, perhaps worst of all in contemporary terms, irrelevant' by people who, upon examination, have made so many mistakes that one cannot but wonder if they have read the books at all. Other 'experts' expend themselves in fatuous witticisms like 'Faërie-land's answer to *Conan the Barbarian*,' and 'Winnie-the-Pooh posing as an epic.' This, then, is the second riddle.

My principal intention in this book is to tackle the first question, and explore the nature of Tolkien's books and their success. However, I think I can also explain, by the same token, why his critics have failed so miserably to do so. I have not taken on *The Silmarillion* here, by the way. The reason is simple: my priority is Tolkien's meaning and impact in the contemporary world, and there is no doubting that that stems almost entirely from *The Lord of the Rings* and, as a kind of introduction, *The Hobbit*. These are his works to which the public has responded, and still does.

My goal means addressing contemporary conditions — cultural, social and political — and readers; and, as far as seems relevant, Tolkien's own character and intentions. But I try to do so while respecting the books' internal integrity; that is, without the single-minded reductionism that sees everything in such a story as 'representing' something else, in line with a predetermined interpretive program around class, or gender, or the unconscious.

The kind of literature which might be said to describe an important part of Tolkien's work, fairy-tales, has been subjected to Freudian, feminist, structuralist, Jungian, anthroposophical and Marxist interpretations in just this way. And they have frequently resulted in some real insights. But too often, the price is a depressing nothing-buttery. Every other dimension of the story is ignored, while the meaning of the whole is tacitly assumed to be exhausted. The spirit-to-letter ratio of these accounts is so low that unlike the stories themselves, they are difficult and dispiriting to read. And behind it all lies a woeful blindness to the power, here and now, of the myths and folk- and fairy-tales themselves.

One tiny example, out of a multitude: it has been asserted (with a degree of seriousness which is hard to determine) that *The Hobbit* represents an alliance between the lower-middle class (Bilbo) and skilled workers, especially working-class miners (the dwarves), in order to overcome a parasitic capitalist exploiter who 'lives off the hard work of small people and accumulates wealth without being able to appreciate its value' (the dragon). This is genuinely interesting, as well as enjoyable; but it says at least as much about Marxism as a fairy-tale as it does about *The Hobbit*, and hardly exhausts either.

I have tried hard to avoid such a practice. It seems to me that every meaningful human discourse has a subjective side as well as an objective one. Relations between the two are complex — for example, the 'inside' can be larger than the 'outside' — and neither (usually the former) can be reduced to or derived from the other without doing irreparable harm to the whole.

For example, seen from the outside, Tolkien's Middle-earth derives from the pagan Norse world-view, plus his knowledge and love of Anglo-Saxon history (Rohan) and medievalism (Gondor), and of trees (all the various forests). One can add to this Tolkien's memories of pre-war rural middle England (the Shire), and of the trenches of World War I, and so on. The result is a complex but ultimately tightly determined and defined place. But for the sym-

pathetic reader, it is not like that at all. He or she stands in an endless dark, damp forest with the light failing; or in a village pub in
multiracial company which ranges from the oddly familiar to the
distinctly odd; or at the foot of mountains which rear ever higher
until stretching out of sight in the unguessable distance. It is effectively unbounded, either in extent or variety.

Any analysis which recognizes only the first world as important, and dismisses or belittles the second, commits the violence
of reductionism. And there is another reason for caution. That
is Tolkien's own warning against an allegorical or purely topical
reading of his story, in which elements receive a literal or one-to-
one interpretation. As he explains in the Foreword to *The Lord of
the Rings*, 'I cordially dislike allegory in all its manifestations, and
always have done so since I grew old and wary enough to detect its
presence. I much prefer history, true or feigned, with its varied applicability to the thought and experience of readers.' Quite so; not
only is allegory unattractively didactic (at best) and bullying (at
worst), but Tolkien is trying here to protect what he had worked
so hard to create, namely a book that is *non*-allegorical. And wisely
so, as that is one of the reasons it has lasted, and continues to find
new generations of readers with their own concerns. For as Tolkien also noted, 'That there is no allegory does not, of course, say
there is no applicability. There always is.' My book precisely concerns the applicability of his work; it is not really about how it
came to be written, or about the man who wrote it.

In any case, I have too much respect for Tolkien's work, in all
its richness, to sacrifice it on the altar of theory. And I have bene-
fitted from some excellent warnings whenever tempted. There is
Gandalf's, of course: 'he that breaks a thing to find out what it is
has left the path of wisdom.' But also T. A. Shippey's: 'Adventure
in Middle-earth embodies a modern meaning, but does not exist
to propagate it.' This seems to me to put the matter perfectly,
along with the shrewd words of Max Luthi: 'Everything external,
not just in literature but also in reality, can be or become a symbol.

It is, however, still itself as well, not only in reality but also in literature.'

So *The Hobbit* and *The Lord of the Rings* are first and foremost, as Tolkien claimed, stories; and ones written by a master storyteller. This is already important for understanding both Tolkien's popular success and his critical slating. Philip Pullman, upon winning the Carnegie Medal for children's fiction in July 1996, put it perfectly: 'in adult literary fiction, stories are there on sufferance. Other things are felt to be more important: technique, style, literary knowingness. Adult readers who do deal in straightforward stories find themselves sidelined into a genre such as crime or science fiction, where no-one expects literary craftsmanship.' Or children's books, which *The Lord of the Rings* is frequently misrepresented as being; or fairy-tales, one of its principal inspirations. 'But,' Pullman continued, 'stories are vital. Stories never fail us because, as Isaac Bashevis Singer says, "events never grow stale." There's more wisdom in a story than in volumes of philosophy.' Most present-day writers, however, are highly anxious to be seen as Grown-Ups. They therefore 'take up their stories as if with a pair of tongs. They're embarrassed by them. If they could write novels without stories in them, they would. Sometimes they do.' Thus the hunger for stories that's there in young and adult alike is unmet, and goes by default to Disney, Hollywood and schlock TV, who are happy to oblige.

As stories, Tolkien's language and style are therefore important. But these have already been tackled in a way I could not better. And imponderables abound. The single greatest obstacle to appreciating Tolkien's work is sheer literary snobbery. But almost equally important is a capacity and liking for imagination, as opposed to a doctrinaire cast of mind. (It may be something like a musical sense.)

Personally, like Hugh Brogan, I find Tolkien's writing 'capable of humour, irony, tragedy, and fast narrative, with only occasional lapses into cardboard grandiloquence.' But even if everyone else agreed, this *alone* would not suffice to explain his appeal. To do so we must turn to their content, and ask: why these particular and apparently rather peculiar stories? For example, how many other world bestsellers are almost entirely devoid of sex? (Except possibly the Bible — a debatable point.) Here, of course, some theory becomes indispensable. So my critical practice, however unsatisfactory it may be (in theory), is to bounce back and forth between the inside and outside of Middle-earth, looking for relations, connections and patterns. In so doing, I have used anything that seems to help, including my own personal and 'subjective' reactions.

My chief concern, as I've said, is the meaning of the work rather than its author. Of course, there is a relationship between the two. But this too is highly complex, and the one does not follow simply from the other. The significance of the work is neither entirely determined nor limited by the life and times that produced it. And as Tolkien himself reminds us of fairy-stories, 'when we have done all that research . . . can do . . . there remains still a point too often forgotten: that is the effect produced *now* by these old things in the stories as they are.' That effect — and only in so far as they are significant for it, Tolkien's sources, influences and so on — is what interests me.

It is boring and pointless to spill ink on whether Tolkien was 'reactionary' or not. Nor can the work itself be pigeonholed in such a ridiculously simplistic way; its meaning is not forever fixed, but rather whatever it presents itself as, in ways that cannot be pre-determined. Indeed, I am going to argue that *The Lord of the Rings* has a life of its own to an extent far exceeding what Tolkien himself expected or could have anticipated. That life is integral to understanding its enduring appeal.

## Postmodernity in Middle-earth

I have derived aid and support from postmodernist theories of meaning and reading that probably would have inspired mixed feelings in Tolkien himself. These offer the starting-point that meaning is tied to shared linguistic and cultural understandings, on the one hand — so that not anything goes — yet meanings are always open, in principle, to re-interpretation along new and different lines, including ones unsuspected by the author. Tolkien can hardly have known when he was writing, for example, that the 1960s were around the corner, and would take up his books with such enthusiasm.

In a way, I myself am another example in this context. Tolkien was a deeply conservative (with a small c) English Roman Catholic with a highly specialized scholarly interest in the early Middle Ages. The best label for me, on the other hand, might be 'Radical Eclectic'; I grew up many generations later in mid-Western Canada and the United States, and was deeply influenced by the intellectual, left-libertarian and mystical aspects of the 1960s . . . including *The Lord of the Rings*. Without the relative independence of the text, my abiding love of it would be impossible to understand.

Postmodernism also holds that while every discipline will have its own set of critical standards for assessing good and bad work, such standards cannot be grounded in any kind of indisputable foundations or ultimate objectivity. They 'are' whatever it is agreed that they are, which of course changes and is never unanimous. So although I have tried to be rigorous and coherent, I make no apology for occasionally explicitly including myself. That is better than pretending to have a total overview from a standpoint that is wholly outside its subject-matter, and therefore supposedly comprehensive and impartial. The contents of books cannot be separated from the sense that particular readers make of them.

Finally, postmodernism has also influenced my account in another important respect. It suggests that we are now living in a time when the project of modernity is approaching exhaustion. What do I mean by modernity? Basically, a 'world-view' that began in late seventeenth-century Europe, became self-conscious in the eighteenth-century Enlightenment, and was exported all over the world, with supreme self-confidence, in the nineteenth. It culminated in the massive attempts at material and social engineering of our own day. Modernity is thus characterized by the combination of modern science, a global capitalist economy, and the political power of the nation-state.

All of these things are now controversial. They used to be justified by the 'grand narratives' of modernity — secularized versions of divine revelation, which were supposed to supply essentially complete accounts of our progress towards the realization of the truth (as laid down by Marx, or Freud, or Darwin). But these no longer command widespread respect or assent. There have been too many broken promises, and too many terrible 'successes': the gulags of universal liberation through class struggle, modern science's showcases at Hiroshima and Chernobyl, and the ongoing holocaust of the natural world at the behest of rational economic development. And while I am as grateful as anyone for the benefits of modernity, and wish to throw out no babies with the bathwater, it is impossible now to avoid the fact that the costs have been horrendous, and are, unlike the benefits, increasing.

Modernity carries on, of course. The power of the state still extends to doing whatever it likes to its (willing or unwilling) citizens, restrained here and there only by the fragile conventions of representative democracy. The development of a superstate ideal in Europe has added further to the load. The highly mixed blessings of 'free' trade are forced on to weaker countries by stronger through GATT and other menacing acronyms. Scientists, following the logic of 'pure knowledge' but backed by big business, are careering ahead with genetic engineering and biotechnology. And when state, science and capital all get together, the result is what Lewis Mumford called 'the Megamachine.' Thus, the same people who brought you nuclear energy, agribusiness and the drug and chemical industries are now pursuing the fantastic corporate profits promised by patenting and selling life itself, under the protection of international law. What price a 'life-form'?

What has changed, with postmodernity, is simply the widespread appearance of questions about the legitimacy and desirability of all this — together with unsettling new reasons and theories for such questions. And people *do* have questions — more people, with more and deeper fears and worries, than perhaps ever before. Only a fool (or convert, or perhaps employee) would say they are groundless. And one of the things being questioned — not a moment too soon — is the value of the kind of deranged, totalizing rationality, epitomized but by no means restricted to modern science, that produces disenchantment. To quote Zygmunt Bauman, postmodernity, above all,

> can be seen as restoring to the world what modernity, presumptuously, had taken away; as a re-enchantment of the world that modernity had tried hard to disenchant. . . . The war against mystery and magic was for modernity the war of liberation leading to the declaration of reason's independence . . . world had to be de-spiritualized, de-animated: denied the capacity of subject. . . . It is against such a disenchanted world that the postmodern re-enchantment is aimed.

## Middle-earth in Postmodernity

I believe Tolkien's books speak to precisely these conditions. Drawing on the power of ancient Indo-European myth, they invite the reader into a compelling and remarkably complete premodern world, saturated with corresponding earlier values, which therefore feels something like a lost home — and by the same token, offers hope for its recovery. They are just the values whose jeopardy we most now feel: relationships with each other, and nature, and (for want of a better word) the spirit, which have not been stripped of personal integrity and responsibility and decanted into a soulless calculus of profit-and-loss; and practical-ethical wisdom, which no amount of economic or technological 'progress' will ever be able to replace. As John Ruskin wonderfully asserted, in the face of Victorian materialist triumphalism in full flood:

> To watch the corn grow, and the blossoms set; to draw hard breath over ploughshare or spade; to read, to think, to love, to hope, to pray — these are the things that make men happy; they have always had the power of doing these, they never *will* have the power to do more. The world's prosperity or adversity depends upon our knowing and teaching these few things: but upon iron, or glass, or electricity, or steam, in no wise.

But as we begin *The Lord of the Rings*, this is exactly the world that is under severe threat from those who worship pure power, and are its slaves: the technological and instrumental power embodied in Sauron (after whom the book itself is named, after all), and the epitome of modernism gone mad. We thus find ourselves reading a story about ourselves, about our own world. That is one reason why so many readers have taken it so to heart.

This analysis has recently found remarkable confirmation. As Bauman also observed, 'people who celebrate the collapse of communism, as I do, celebrate more than that without always knowing it. They celebrate the end of modernity actually, because what collapsed was the most decisive attempt to make modernity work; and it failed. It failed as blatantly as the attempt was blatant.' Now, *The Hobbit* and *The Lord of the Rings* were already underground 'cult' classics in the USSR, Czechoslovakia, Poland and Hungary before 1989. Since then they have boomed there in a way reminiscent of the late 1960s in the West. But the exhilaration of liberation is already fast succumbing to the discovery that 'free market' capitalism, as such, is simply a more efficient version of the same economic logic as its former state form. I fear Tolkien will have no shortage of newly disillusioned readers there.

Tolkien himself, of course, was deeply hostile to modernity, root and branch — capitalism (especially industrialism), unrestrained science, and state power alike. For him, they were idols whose worship had resulted, in our century, in the most efficient ever devastation of both nature and humanity alike. He once remarked that 'I would arrest anybody who uses the word State (in any sense other than the inanimate realm of England and its inhabitants, a thing that has neither power, rights nor mind) . . .' And he described the detonation of the atom bomb in 1945 as 'the utter folly of these lunatic physicists.' But that is not a very original observation, and neither so interesting nor significant as what has become of his *anti*-modernism, lovingly and skilfully embodied in a literary artefact, in *post*modern times. As he himself put it, 'it is the particular use in a particular situation of any motive whether invented, deliberately borrowed, or unconsciously remembered, that is the most interesting thing to consider.'

Now, it is perfectly possible to imagine Tolkien's books 'being' truly reactionary: racist, nationalist, etc. I contend, however, that as it happens — as things have actually turned out — his implicit diagnosis of modernity was prescient; and his version of an alternative, progressive. That is, in the context of global modernization and the resistance to it, his stories have become an animating and inspiring new myth. It joins up with a growing contemporary sense, represented in postmodernism, of history's sheer contingency: a liberating perception that things might have been different, and therefore *could* be different now. It suggests that just as there was life before modernity, so there can be after it.

In short, Tolkien's books are certainly nostalgic, but it is an emotionally empowering nostalgia, not a crippling one. (The word itself means just 'homesickness.') One contemporary writer, Fraser Harrison, goes straight to the heart of the matter: 'While it is easy to scoff at the whimsicality and commercialism of rural nostalgia, it is also vital to acknowledge that this reaching-out to the countryside is an expression, however distorted, of a healthy desire to find some sense of meaning and relief in a world that seems increasingly bent on mindless annihilation.' Accordingly, says Harrison, 'it becomes meaningful to talk of "radical nostalgia".'

Only those who cling to the modernist myth of a singular universal truth (as opposed to myth and story and indeed interpretation as such) which is somehow directly accessible to those with the 'correct' understanding — only such people will look at Tolkien's glorious tree and see, to use an apt image of William Blake's, nothing more than 'a Green thing that stands in the way.' To the modernist, the choice is between truth and myth (or falsehood), whereas the postmodernist, giving up the pretence of a direct line to the Truth, sees the choice as between different truths; or to put it another way, between myths and stories that are creative and liberating, and those that are destructive and debilitating. As Tolkien put it, 'History often resembles "Myth," because they are both ultimately of the same stuff.'

Ironically, therefore, it is Tolkien's critics who have been overtaken by events. Behind their instinctive antagonism lies an uncomfortable sense that here is a coherent fictional critique and an alternative, in every major respect, to the exhausted myth of modernity which has so far underwritten their own professional status; and worse, it is a popular one! Not for the first time, those who claim to know better than and even speak for 'the people' are lagging behind them.

## Three Worlds in One

I have said that Tolkien's literary creation presents a remarkably complete alternative world, or rather, alternative version of our world. I myself only realized its depth and complexity when I tackled it in a spirit of determined but non-reductionist analysis. There are almost no threads that can be tugged without them leading on to others, almost indefinitely. But I found I could make sense of most of it in terms of three domains, each one nesting within a larger: the social ('the Shire'), the natural ('Middle-earth'), and the spiritual ('the Sea'). I was encouraged in this by Tolkien's own remark in his superb essay on the subject, that 'fairy-stories as a whole have three faces: the Mystical towards the Supernatural; the Magical towards Nature; and the Mirror of scorn and pity towards Man.'

Thus, *The Lord of the Rings* begins and ends with the hobbits, in the Shire. This is the social, cultural and political world. It includes such things as the hobbits' strong sense of community, their decentralized parish or municipal democracy, their bioregionalism (living within an area defined by its natural characteristics, and within its limits), and their enduring love of, and feeling for, place. In all these respects, the ultimate contrast is with the brutal universalism and centralized efficiency of totalitarian Mordor.

Now this sphere is indeed crucial, but it nests within a larger and weightier world, just as the Shire itself does: namely the extraordinarily varied and detailed natural world of Middle-earth. Note that this therefore *includes* the human world. Tolkien plainly had a profound feeling for nature, and perhaps especially its flora; his love of trees shines through everywhere. The sense in *The Lord of the Rings* of a tragically endangered natural world, savaged by human greed and stupidity in every corner of the globe, is confirmed for us in every daily newspaper. But this 'nature' is neither romantic nor abstract. There are plenty of dangerous wild places in Middle-earth; but they are all, like their blessed counterparts, very specific places. Indeed, Tolkien's attention to 'local distinctiveness' is one of the most striking things about his books. It contributes greatly to the uncanny feeling, shared by many of his readers, of actually having been there, and knowing it from the inside, rather than simply having read about it — the sensation, as one put it, of 'actually walking, running, fighting and breathing in Middle-earth.'

Above all, Tolkien's is no add-on environmentalism. It suggests rather that whatever their differences, humans *share* with other living beings a profound common interest in life, and whatever aids life. Thus Middle-earth's most distinctive places defy the separation, so beloved of modernist scientific reason, into 'human or social and therefore conscious subjects' and 'natural and therefore inert objects.' They are both: the places themselves are animate subjects with distinct personalities, while the peoples are inextricably in and of their natural and geographical locales: the Elves and 'their' woods and forests, the Dwarves and mountains, hobbits and the domesticated nature of field and garden. And some of the most beautiful places in Middle-earth are so, in large part, because they are loved by the people who share them. Tolkien's prescient ecologism is therefore radical, in the modern sense as well as the old one of a return to roots. It anticipates, in many ways, both 'social' and 'deep' ecology, and retraces a premodern way of understanding the world which is still that of surviving indigenous tribal peoples. Time is running out for the rest of us to re-learn it.

Following this up, I then found myself at the edge of the second circle too. In Tolkien's terms, I had been brought up short by the Sea. This third sphere proved to be the most encompassing of all: an ethics rooted, so to speak, in spiritual values as symbolized by the Sea. Here we shall discover the way Tolkien deals with the problem of spirit in a secular age; a problem with, as Salman Rushdie once put it, a God-shaped hole in it, but equally, with some very good reasons to resist any simple re-insertion of God. Indeed, despite his personal religious convictions, Tolkien was acutely aware of writing in and for a divided post-Christian audience — just as one of his heroes, the author of *Beowulf,* had been at the beginning of the same era. His book therefore makes no explicit references to any organized religion at all, and (unlike those of his sometime friend C. S. Lewis) offers no hostages to a religiously allegorical interpretation.

As we shall see, the spiritual world of Middle-earth is a rich and complex one. It contains both a polytheist-cum-animist cosmology of 'natural magic' and a Christian (but non-sectarian) ethic of humility and compassion. Tolkien clearly felt that both are now needed. The 'war against mystery and magic' by modernity urgently requires a re-enchantment of the world, which a sense of Earth-mysteries is much better-placed to offer than a single transcendent deity. (As Gregory Bateson once remarked, when the loss of a sense of divine immanence in nature is combined with an advanced technology, 'your chance of survival will be that of a snowball in hell.') But the Christian dimension of humility and ultimate dependency, exemplified by Frodo, is the best answer to modernity's savage pride in the efficiency, and self-sufficiency, of its own reason. Rising above the dogmas of his own religious upbringing, Tolkien has thus made it possible for his readers to unselfconsciously combine Christian ethics and a neo-pagan reverence for nature, together with (no less important) a liberal humanist respect for the small, precarious and apparently mundane. This is a fusion that couldn't be more relevant to resisting the immense and impersonal forces of runaway modernity.

In what follows, I shall be looking at the social, natural and spiritual aspects of Tolkien's world in turn, and their crucial overlap. That is where their heart is to be found, and any meaning found in or derived from his work must embody all three concerns to be considered essential. Taken together, they comprise the whole implicit project of his literary mythology, and a remedy for pathological modernity in a nutshell: namely, *the resacralization (or re-enchantment) of experienced and living nature, including human nature, in the local cultural idiom.* I am not at all suggesting, of course, that were everyone to read Tolkien everything would be fine; just that his books have something, however small, to contribute to a collective healing process.

More modestly still, critical recognition of this project and contributions to it like Tolkien's might help restore a pathologically, almost terminally jaded critical community. To quote Ihab Hassan, 'I do not know how to give literature or theory or criticism a new hold on the world, except to remythify the imagination, at least locally, and bring back the reign of wonder into our lives.' Such a response to modernity is no mere escapist sentimentality. In fact, as we ought to know at the end of this bloody century, it is not the sleep of reason that produces monsters, but sleepless reason. Tolkien realized this, with implications I shall discuss in relation to 'mythopoeic' literature:

> Fantasy is a natural human activity. It certainly does not destroy or even insult Reason . . . On the contrary. The keener and the clearer is the reason, the better fantasy it will make. If men were ever in a state in which they did not want to know or could not perceive truth (facts or evidence), then Fantasy would languish. . . . For creative Fantasy is founded upon the hard recognition that things are so in the world as it appears under the sun; on a recognition of fact, but not a slavery to it.

## A Mythology for England?

For Tolkien himself, of course, and for English readers, the native cultural idiom happens to be an English one. Part of Tolkien's ambition was 'to restore to the English an epic tradition and present them with a mythology of their own' — something that he felt was lacking in their national literature. (The Arthurian myth-cycle was, he felt, powerful but 'imperfectly naturalized': more British, that is, Celtic, than English, with its faërie 'too lavish,' and in addition — what struck Tolkien, for reasons we shall explore, as 'fatal' — it explicitly contains Christianity.) Tolkien was not the only one to feel such a lack. In 1910, E. M. Forster wrote: 'Why has not England a great mythology? Our folklore has never advanced beyond daintiness, and the greater melodies about our countryside have all issued through the pipes of Greece. Deep and true as the native imagination can be, it seems to have failed here. It has stopped with the witches and fairies . . .'

Tolkien blamed this on the brutality of the Norman occupation beginning in 1066, and not without reason. It was a savage assault on a relatively peaceful land, which eventually left one person in ten there dead from war or starvation. It also imposed a new phenomenon on the British Isles: a foreign and highly centralized ruling class, including secular, ecclesiastical and educational élites. The new Norman archbishop, bishops and abbots regarded their Anglo-Saxon ecclesiastical predecessors as *rudes et idiotas* (uncouth and illiterate), dropped the worship of many pre-conquest saints and even destroyed some of their shrines. Education now demanded Latin, and 'culture,' as well as power, French; for as long as two hundred years later, the nobility still did not speak the native tongue. And Tolkien's modern critics today are the heirs of precisely the same caste, almost as divorced now from the common reader as their forebears were from the common people, and no less lofty in attitude.

For our purposes here, however, the point is the way biography can transmute, through art, into contemporary relevance. For Tolkien's deep dislike of the Norman virtues of bureaucracy, efficiency and rationalization, as it manifests itself in *The Lord of the Rings*, provides the contemporary reader with an instant 'recognition' of the global modernization which the Normans, as it happens, anticipated in these important respects.

But Englishness is not inscribed in the text. This is something I finally realized after talking to Russian and Irish and Italian readers, and discovering that each one had found in the hobbits an accessible native tradition, centred on a 'small,' simple and rural people — and self — with which to begin, and end renewed.

I am not just talking about long-vanished peasants, either. I know one man living within a few minutes of both the diabolical London motorway ring-road and Heathrow Airport whom Farmer Maggot could have been modelled on: 'There's earth under his old feet, and clay on his fingers; wisdom in his bones, and both his eyes are open.' Of course, he was living there before these monstrosities appeared; but they haven't driven him out. Such people in such places may have gone to ground, but they're still around, and there are even some younger ones coming up. As the Donga 'tribe' (named after the ancient trackways on Twyford Down, Hampshire) sing, 'We are the old people, we are the new people, we are the same people, stronger than before.'

Nor are *The Hobbit* and *The Lord of the Rings* ethnocentrically limited to northwestern Europe — even though the qualities of their peoples, lands, seasons, the very air belong to that part of the world. The reason is another of Tolkien's master-strokes. The anthropologist Virginia Luling has pointed out that he presents us with a northwestern Europe, the home and heartland of the industrial revolution, as a place where it has never happened; and by the same token, with the birthplace of colonialism and imperialism as an unstained 'Fourth World' of indigenous tribes. Accordingly, the cultures of Middle-earth's peoples are pre-modern or 'traditional,' and indeed pre-Christian, while their religions and mythologies are animist, polytheist and shamanist. But Tolkien's choice of a 'Norse' mythology for his tale as a whole, over the usual Graeco-Roman one, situates his story still more precisely. (It also effectively bypasses all the élite critical apparatus of Greek and Latin references which many ordinary readers might find either boring or alienating.)

In fact, the only place in Middle-earth which is industrial, imperialistic, and possessed of an all-powerful state is Mordor — admittedly the most powerful force of all, as such, but essentially an alien invader (as Sauron originally was) rather than a native. Tolkien's Middle-earth is thus a Europe, as Luling puts it, that has never been 'Europeanized,' or, what amounts to the same thing, 'modernized.' And the story of *The Lord of the Rings* — as reflected in its very title — is about the resistance to just that. The potential relevance of these books consequently opens out not only to anyone living *in* 'the West,' but to anyone affected *by* it; which is to say, nearly everyone anywhere.

## A Great Book?

We shall also consider *The Lord of the Rings* as literature. That involves considering why Tolkien chose 'fantasy,' with its affinities with fairy-tale and myth, as the appropriate form and strategy; and why the wisdom of that choice has been so roundly confirmed by readers, although ignored or condemned by critics. There is also the question of comparable books. I shall suggest that there are indeed a few other works of literary myth, or 'mythopoeic' fiction, which also reveal its true power, feed the soul, and escape the modernist critical compass. There are others, also apparently 'fantasy,' which are completely different. Here I am obliged to be unkind to some sacred cows: from the pernicious productions of The Walt Disney Company, and pseudo-fairy-tales like *The Wizard of Oz*, to some of the authors and critics now canonized by literary feminism.

Given how unavoidably subjective and personal it must be, compiling lists of 'great' books is a game we can all play. I have no doubt that *The Lord of the Rings* is one of the greatest works of twentieth-century literature, even if not always for purely 'literary' reasons. But I am not too concerned to persuade the reader to agree; just to realize that it is fully deserving of affection and respect, and even some passionate attention. Written with love, learning, skill and sacrifice, it is a cry (as someone once said of religion) from 'the heart of a heartless world, the soul of soulless conditions,' but also something more. It offers not an 'escape' from our world, this world, but hope for its future.

# THE SHIRE: CULTURE,
# SOCIETY AND POLITICS

*It is as neighbours, full of ineradicable
prejudices, that we must love each other, and
not as fortuitously 'separated brethren.'*

HOBBITS, ACCORDING TO Tolkien, were more frequent 'long
ago in the quiet of the world . . .' They 'love peace and quiet and
good tilled earth: a well-ordered and well-farmed countryside was
their favourite haunt. They do not and did not understand or like
machines more complicated than a forge-bellows, a water-mill, or
a hand-loom. . . . Their faces were as a rule good-natured rather
than beautiful, broad, bright-eyed, red-cheeked, with mouths apt
to laughter, and to eating and drinking.' They thought of themselves as 'plain quiet folk' with 'no use for adventures. Nasty disturbing uncomfortable things! Make you late for dinner!' 'Nonetheless,' their chronicler notes, 'ease and peace had left this people
still curiously tough. They were, if it came to it, difficult to daunt
or to kill . . .' In other words, they manifested 'the notorious Anglo-hobbitic inability to know when they're beaten.'

Hobbits were also inclined 'to joke about serious things,' and
'say less than they mean.' Indeed, they 'will sit on the edge of ruin
and discuss the pleasures of the table, or the small doings of their

fathers, grandfathers, and great-grandfathers, and remoter cousins to the ninth degree, if you encourage them with undue patience.' Similarly, they preferred speeches that were 'short and obvious,' and 'liked to have books filled with things that they already knew, set out fair and square with no contradictions.' They were 'a bit suspicious . . . of anything out of the way — uncanny, if you understand me.' It wasn't difficult to acquire a reputation for peculiarity in the Shire.

But as Tolkien notes, in addition to their wealth 'Bilbo and Frodo Baggins were as bachelors very exceptional, as they were also in many other ways, such as their friendship with Elves.' The nephew of 'mad Baggins,' as he eventually became known, Frodo was something of an aesthete and intellectual, who, 'to the amazement of sensible folk . . . was sometimes seen far from home walking in the hills and woods under the starlight.' None of this was usual among their peers, and Sam the gardener, although recently and exceptionally lettered, was a more typical hobbit than his fellow Companions — or as Tolkien put it, 'the genuine hobbit.'

Like some readers, Tolkien himself sometimes found Sam, as he wrote:

> very 'trying.' He is a more representative hobbit than any others that we have to see much of; and he has consequently a stronger ingredient of that quality which even some hobbits found at times hard to bear: a vulgarity — by which I do not mean a mere 'down-to-earthiness' — a mental myopia which is proud of itself, a smugness (in varying degrees) and cocksureness, and a readiness to measure and sum up all things from a limited experience, largely enshrined in sententious traditional 'wisdom.' . . . Imagine Sam without his education by Bilbo and his fascination with things Elvish!

Even with this kind of conservative peer pressure, however, your behaviour had to be extreme to land you in any real trouble, for the Shire at this time had hardly any government: 'Families for the most part managed their own affairs. . . . The only real official in the Shire at this date was the Mayor of Michel Delving,' and 'almost his only duty was to preside at banquets . . .' Otherwise there were only hereditary heads of clans, plus a Postmaster and First Shirriff — the latter less for Inside Work than 'to see that Outsiders of any kind, great or small, did not make themselves a nuisance.'

## Englishness

Now it doesn't take any great perceptiveness to see in 'these charming, absurd, helpless' (and not-so-helpless) hobbits a self-portrait of the English, something which Tolkien admitted: '"The Shire" is based on rural England and not any other country in the world,' and more specifically the West Midlands: Hobbiton 'is in fact more or less a Warwickshire village of about the period of the Diamond Jubilee' (i.e. 1897).

Compare the portrait by George Orwell writing in 1940, and one still instantly recognizable, albeit sadly altered in some respects, of a conservative people neither artistically nor intellectually inclined, though with 'a certain power of acting without thought;' taciturn, preferring tacit understandings to formal explication; endowed with a love of flowers and animals, valuing privateness and the liberty of the individual, and respecting constitutionalism and legality; not puritanical and without definite religious belief, but strangely gentle (and this has changed most, especially during the 1980s), with a hatred of war and militarism that coexists with a strong unconscious patriotism. Orwell summed up English society as 'a strange mixture of reality and illusion, democracy and privilege, humbug and decency.'

True, these attributes are inextricably mingled with ones that the English have wanted to find in the mirror; nor are they eternal and immutable. Because this image partakes of a national pastoral fantasy, however, it does not follow that it has no reality. A social or literary criticism that is afraid to admit the relative truth of clichés and stereotypes is hamstrung from the start. Also, it is worth noting that Tolkien's portrait is not altogether a flattering one; it includes greed, small-minded parochialism and philistinism, at least — even if Frodo, Sam and the other hobbits of his story were able to rise above these regrettable characteristics of the English bourgeoisie.

However, although Tolkien drew on the tiny corner of the world that is the West Midlands of England, readers from virtually everywhere else in the world connect the hobbits with a rustic people of their own, relatively untouched by modernity — if not still actually existing, then from the alternative reality of folk- and fairy-tale. Doubtless this has been made possible by setting his books in a place that, while it feels like N.W. Europe, is made strange and wonderful by its imaginary time. Otherwise, I have no doubt, they would have suffered from the same limitations of time and place as Kipling's *Puck of Pook's Hill* and G. K. Chesterton's poems, however wonderful these otherwise may be. Tolkien's tale, in contrast, has probably achieved as close to universality as is given to art.

## Country Folk

The hobbits are recognizably modern in important respects, especially in their bourgeois and anti-heroic tenor. Thus, one famous hobbit, when asked by a large eagle, 'What is finer than flying?,' only allowed his native tact, and caution, to overrule suggesting 'A warm bath and late breakfast on the lawn afterwards.' As several commentators have noticed, it is crucial that Bilbo and Frodo *be* modern, in order to 'accommodate modernity without surrendering to it,' by mediating between ourselves and the ancient and foreign world they inhabit. But in other ways, the hobbits have much older roots. They remind us of 'the archetypal pre-Industrial Revolution English yeomen with simple needs, simple goals, and a common-sense approach to life,' and also of the English before their defeat in 1066, when the 'Norman Yoke' imposed central-ized autocratic government, a foreign language and an alien cul-tural tradition.

The bucolic hobbits also clearly fall within the long tradi-tion in English letters of nostalgic pastoralism, celebrating a time 'when there was less noise and more green.' As Martin J. Weiner notes, 'Idealization of the countryside has a long history in Brit-ain.' It extends from Tennyson's mid-Victorian *English Idylls* and William Morris's 'fair green garden of Northern Europe,' through the rural essays of Richard Jefferies and the Poet Laure-ate Alfred Austin's *Haunts of Ancient Peace* (1902) — which could easily be the title of a song by Van Morrison today — to Kipling's 'Our England is a garden,' and George Sturt listening to his gar-dener (note), 'in whose quiet voice,' he felt, 'I am privileged to hear the natural fluent, unconscious talk, as it goes on over the face of the country, of the English race.' In short, a deep cultural gulf had opened between England's southern and rural 'green and pleasant land' and her northern and industrial 'dark satanic mills'; or as Weiner puts it, with unintentional aptness, 'The power of the machine was invading and blighting the Shire.'

The irony is, of course, that since 1851 over half the popula-tion on this island has lived in towns, and by then England was al-ready the world's first urban nation. Thus, as Weiner writes, 'The less practically important rural England became, the more easily could it come to stand simply for an alternative and complemen-tary set of values, a psychic balance wheel.' But few things are that simple, and when applied to Tolkien, such glib simplification has led to a great deal of misunderstanding. The related charges com-monly laid at Tolkien's door are several, and severe. They are also almost entirely mistaken, so I shall use them to arrive at the truth of the matter.

## Nation and Class

One of the first critics to attack Tolkien was Catherine Stimpson, in 1969. 'An incorrigible nationalist,' she wrote, Tolkien 'celebrates the English bourgeois pastoral idyll. Its characters, tranquil and well fed, live best in placid, philistine, provincial rural cosiness.'

Now it is true that the hobbits (excepting Bilbo and Frodo, and perhaps Sam . . . and Merry and Pippin) would indeed have preferred to live quiet rural lives, if they could have. Unfortunately for them, and Stimpson's point, there is much more to Middle-earth than the Shire. By the same token, any degree of English nationalism that the hobbits represent is highly qualified. Tolkien himself pointed out that 'hobbits are not a Utopian vision, or recommended as an ideal in their own or any age. They, as all peoples and their situations, are an historical accident — as the Elves point out to Frodo — and an impermanent one in the long view.' It is also possible, as Jonathan Bate suggests, to draw a distinction between love of the local land, on the one hand, and patriotic love of the fatherland on the other. In *The Lord of the Rings*, the lovingly detailed specificities of the natural world — which include but far outrun those of the Shire — far exceed any kind of abstract nationalism.

Stimpson also accuses Tolkien of 'class snobbery' — that is, the lord of the manor's disdain for commoners, and, by extension, the working class. Well, in *The Hobbit*, perhaps; but only zealous detectors of orcism and trollism would ignore its other virtues, such as any quality as a story. And its hero, if no peasant, is plainly no lord. But with *The Lord of the Rings* — if this charge means anything worse than a sort of chivalrous paternalism, appropriate to someone growing up at the turn of this century, which now looks dated — then it fails.

There is certainly class awareness. But the idioms of Tolkien's various hobbits only correspond to their social classes in the same way as do those of contemporary humans. The accent and idiom of Sam (arguably the real hero of the book) and most other hobbits are those of a rural peasantry, while those of Frodo, Bilbo and their close friends range through the middle classes. Or take Orcs; their distinguishing characteristics are a love of machines and loud noises (especially explosions), waste, vandalism and destruction for its own sake; also, they alone torture and kill for fun. Their language, accordingly, is 'at all times full of hate and anger,' and composed of 'brutal jargons, scarcely sufficient even for their own needs, unless it were for curses and abuse.' In the Third Age, 'Orcs and Trolls spoke as they would, without love of words or things; and their language was actually more degraded and filthy,' writes Tolkien, 'than I have shown it.' As he adds, too truly, 'Much the same sort of talk can still be heard among the orc-minded; dreary and repetitive with hatred and contempt . . .'

But Orc speech is not all the same; there are at least three kinds, and none is necessarily 'working-class.' And it can be found today among members of any social class; nor is money a bar. In fact, virtually all of Tolkien's major villains — Smaug, Saruman, the Lord of the Nazgûl, and presumably Sauron too — speak in unmistakably posh tones. After all, the orc-minded are mere servants of Mordor; its contemporary masters (or rather, master-servants) much more resemble the Nazgûl, although today they probably wear expensive suits and ride private jets rather than quasi-pterodactyls. And although many fewer than Orcs (who knows? perhaps there are exactly nine), they are infinitely more powerful, and to be feared.

There is also the obvious and fundamental fact of *The Lord of the Rings* as a tale of 'the hour of the Shire-folk, when they arise from their quiet fields to shake the towers and counsels of the Great.' Nonetheless, the charge of pandering to social hierarchy has proved durable. Another unpleasant and related accusation sometimes made is racism. Now it is true that Tolkien's evil creatures are frequently 'swart, slant-eyed,' and tend to come from the south ('the cruel Haradrim') and east ('the wild Easterlings') — both threatening directions in Tolkien's 'moral cartography.' It is also true that black — as in Breath, Riders, Hand, Years, Land, Speech — is often a terrible colour, especially when contrasted with Gandalf the White, the White Rider, and so on. But the primary association of black here is with night and darkness, not race. And there are counter-examples: Saruman's sign is a white hand; Aragorn's standard is mostly black; the Black Riders were not actually black, except their outer robes; and the Black Stone of Erech is connected with Aragorn's forebear, Isildur.

Overall, Tolkien is drawing on centuries of such moral valuation, not unrelated to historical experience attached to his chosen setting in order to convey something immediately recognizable in the context of his story. As Kathleen Herbert noticed, Orcs sound very like the first horrified reports in Europe of the invading Huns in the fourth and fifth centuries: 'broad-shouldered, bow-legged, devilishly effective fighters, moving fast, talking a language that sounds like no human speech (probably Turkic) and practising ghastly tortures with great relish.' (Théoden may well have been modelled on Theodoric I, the aged Visigothic king who died leading his warriors in a charge against Attila's Huns in the Battle of Chalons.)

Perhaps the worst you could say is that Tolkien doesn't actually go out of his way to forestall the possibility of a racist interpretation. (I say 'possibility' because it is ridiculous to assume that readers automatically transfer their feelings about Orcs to all the swart or slant-eyed people they encounter in the street.) But as Virginia Luling has pointed out, the appearance of racism is deceptive, 'not only because Tolkien in his non-fictional writing several times repudiated racist ideas, but because . . . in his sub-creation the whole intellectual underpinning of racism is absent.' In any case, such an interpretation, as the story in *The Lord of the Rings* proceeds, would get increasingly harder to maintain — and this relates to another common criticism, also voiced by Stimpson, that Tolkien's characters divide neatly into 'good and evil, nice and nasty.' But as anyone who has really read it could tell you, the initial semi-tribal apportioning of moral probity increasingly breaks down, as evil emerges 'among the kingly Gondorians, the blond Riders of Rohan, the seemingly incorruptible wizards, and even the thoroughly English hobbit-folk of the Shire.' (Incidentally, hobbits appear to be brown-skinned, not white.) By the same token, Frodo, Gollum, Boromir and Denethor all experience intense inner struggles over what the right thing to do is, with widely varying outcomes; and as Le Guin has noted, several major characters have a 'shadow.' In Frodo's case, there are arguably two: Sam and Gollum, who is himself doubled as Gollum/Stinker and Sméagol/Slinker, as Sam calls him.

'If you want to write a tale of this sort,' Tolkien once wrote, 'you must consult your roots, and a man of the North-west of the Old World will set his heart and the action of his tale in an imaginary world of that air, and that situation: with the Shoreless Sea of his innumerable ancestors to the West, and the endless lands (out of which enemies mostly come) to the East.'

Thus, as Clyde Kilby recounts, when Tolkien was asked what lay east and south of Middle-earth, he replied: '"Rhûn is the Elvish word for east. Asia, China, Japan, and all the things which people in the West regard as far away. And south of Harad is Africa, the hot countries." Then Mr. Resnick asked, "That makes Middle-earth Europe, doesn't it?" To which Tolkien replied, "Yes, of course — Northwestern Europe . . . where my imagination comes from".' (In which case, as Tolkien also agreed, Mordor 'would be roughly in the Balkans.')

He reacted sharply to reading a description of Middle-earth as 'Nordic,' however: 'Not *Nordic*, please! A word I personally dislike; it is associated, though of French origin, with racialist theories . . .' He also contested Auden's assertion that for him 'the North is a sacred direction': 'That is not true. The North-west of

Europe, where I (and most of my ancestors) have lived, has my affection, as a man's home should. I love its atmosphere, and know
more of its histories and languages than I do of other parts; but it
is not "sacred," nor does it exhaust my affections.'

It is also striking that the races in Middle-earth are most striking in their variety and autonomy. I suppose that this *could* be seen
as an unhealthy emphasis on 'race'; it seems to me rather an assertion of the wonder of multicultural difference. And given that
most of Middle-earth's peoples are closely tied to a particular geography and ecology, and manage to live there without exploiting
it to the point of destruction, isn't this what is nowadays called
bioregionalism? But no kind of apartheid is involved: one of the
subplots of *The Lord of the Rings* concerns an enduring friendship between members of races traditionally estranged (Gimli and
Legolas), and the most important union in the book, between
Aragorn and Arwen, is an 'interracial' marriage. As usual, the picture is a great deal more complex than the critics, although not
necessarily the public, seem to see.

## A Pastoral Fantasy?

A major stream of hostile Tolkien criticism can be traced back to
Raymond Williams, who fathered British cultural studies, and
called his method 'cultural materialism.' In *The Country and the
City*, Williams noted the 'extraordinary development of country-
based fantasy, from Barrie and Kenneth Grahame through J. C.
Powys and T. H. White and now to Tolkien . . .' and concluded, 'It
is then not only that the real land and its people were falsified; a
traditional and surviving rural England was scribbled over and almost hidden from sight by what is really a suburban and half-educated scrawl.'

Williams has been massively influential. One could produce
many other commentators since who have lambasted pastoralism
in the same way. One writes of 'the ultimate, deeply conservative,

ambition of pastoral' that it 'falsifies the actual relations of non-city communities just as much and for the same reason that it falsifies city communities.' For another, 'The Pastoral allows for a direct opposition to social change, a reactionary clinging to a static present, and an often desperate belief in future improvement.' And it fades away with 'the possibility of social mobility and of economic progress.' (How dated this now sounds, as we face increasingly insurmountable problems as a direct result of 'economic progress'!)

Let us put cultural materialism to the test by seeing how well it applies to Tolkien. According to Williams, 'In Britain, identifiably, there is a precarious but persistent rural-intellectual radicalism: genuinely and actively hostile to industrialism and capitalism; opposed to commercialism and the exploitation of the environment; attached to country ways and feelings, the literature and the lore.' This sounds generous, until you get to the punch-line: 'in every kind of radicalism the moment comes when any critique must choose its bearings, between past and future. . . . We must begin differently: not in the idealizations of one order or another, but in the history to which they are only partial and misleading responses.' By the same token, according to Williams, in our current crises myth and revolution are opposites rather than complementary: we must have 'real history' oriented to a revolutionary future, not 'myth' dreaming of the past.

But this emperor now has no clothes, if indeed he ever had. The mythical 'vs.' the actual, the ideal 'vs.' the real — this is a set of choices which postmodern sensibilities have exposed as cruelly misleading. The 'material' is meaningless except as structured by ideas; conversely, ideas have highly material effects. Revolutions — before, during and after — are saturated with myth. Nor is the political character of traditions and positions inherent and fixed for all time; look how Marxism-Leninism, supposedly 'left-wing,' became crudely authoritarian; or how 'conservative' parties today have become vehicles of sweeping radical change. Williams doesn't even seem to realize that people do not live by factual and physical bread alone, but also by ideas, values and visions of alternatives.

It is not surprising, then, that his treatment of pastoralism terminates in mere abuse of Tolkien's work as, absurdly, 'half-educated' and 'suburban.' Oxford professors may be many things, but they are not yet half-educated; and Tolkien actually complained to his son in 1943 that 'the bigger things get the smaller and duller or flatter the globe gets. It is getting to be all one blasted little provincial suburb.' Nor has Williams noticed that the hobbits' pastoralism is dominated and subverted by other themes. As Gildor said to Frodo, 'it is not your own Shire. . . . Others dwelt here before hobbits were; and others will dwell here again when hobbits are no more. The wide world is all about you: you can fence yourselves in, but you cannot for ever fence it out.' As Merry too admitted, 'It is best to love first what you are fitted to love, I suppose: you must start somewhere and have some roots, and the soil of the Shire is deep. Still there are things deeper and higher; and not a gaffer could tend his garden in what he calls peace but for them, whether he knows about them or not.' *The Lord of the Rings* could thus better be seen as an extended argument that pastoralism alone is not enough — doomed, even: 'The Shire is not a haven, and the burden of the tale is that there are no havens in a world where evil is a reality. If you think you live in one, you are probably naïve like the early Frodo, and certainly vulnerable.'

Perhaps the political problem is the richness and centrality of the natural world in Middle-earth (and not just pastoral nature). But if so, it only serves to confirm that the Left of Williams and his followers remains stuck in a modernist and economistic worldview. Had Marxist socialism accepted William Morris's generous offer to meet halfway (as E. P. Thompson put it), this tragedy never would have happened.

Thompson himself is a good counter-example: Morris's biographer, a passionate critic of economism and class reductionism,

defender of William Blake's *mythos*, and, perhaps not so coinci-
dentally, a passionate gardener. Here, in a catalogue that would
have impressed even Samwise, is Thompson's account of his garden on his fiftieth birthday: 'there is: rasps, strawbs, red, white and
black currants, worcester berries, wineberries, gooseberries, lo-
ganberries, lettuces, radishes, asparagus, tomatoes, globe artichokes, Jerusalem artichokes, marrow, cucumber, broad beans,
peas, runner beans, french beans, rhubarb, cabbage, broccoli, car-
rots, leeks, spring onions, celery, CORN, apples, peaches, nectar-
ines and weeds.'

Thompson is one powerful reminder that in order to be progressive it is not helpful, let alone necessary, to adopt the po-faced
dogma of materialist and rationalist modernism. George Orwell
(also a gardener), is another. In 'Some Thoughts on the Common
Toad' (1946), he asked:

> Is it wicked to take a pleasure in spring? . . . is it politically
> reprehensible, while we are all groaning, or at any rate ought
> to be groaning, under the shackles of the capitalist system, to
> point out that life is frequently more worth living because of
> a blackbird's song, a yellow elm tree in October, or some
> other natural phenomenon which does not cost money and
> does not have what the editors of left-wing newspapers call a
> class angle?

## *Fascist?*

Williams says that nostalgic 'celebrations of a feudal or aristocratic order' embody values that 'spring to the defence of certain kinds of order, certain social hierarchies and moral stabilities,
which have a feudal ring but a more relevant and more dangerous contemporary application . . . in the defence of traditional
property settlements, or in the offensive against democracy in the
name of blood and soil.'

Williams' disciple and biographer Fred Inglis has made the unpleasant implications of this passage explicit in relation to Tolkien,
whose 'schmaltz-*Götterdämmerung*' (he wrote) is such that 'for
once it makes sense to use that much-abused adjective, and call
Tolkien a Fascist.' He later retracted this outrageous slur only to
claim the same thing of *The Lord of the Rings:* 'instead of Nurem-
berg, Frodo's farewell.'

So let us consider the politics (in the narrow sense) of both
Tolkien and Middle-earth. Before doing so, however, I would like
to point out that there is simply no Wagnerian *'Götterdämmerung'*
in *The Lord of the Rings;* 'Victory neither restores an earthly Paradise nor ushers in New Jerusalem.' In addition, Tolkien disliked
Wagner's *Der Ring des Nibelungen,* with which his work has often
been bracketed — 'Both rings were round,' he once snapped, 'and
there the resemblance ceases' — and all the more so for drawing
directly on some of the same mythological material that Wagner
only knew second-hand, and used to such very different ends. (Interestingly, Ragnarok was a relatively late aspect of Germano-
Scandinavian mythology that never caught on in the pagan Anglo-Saxon England that so influenced Tolkien. Even then, it was,
apparently, un-English in its melodrama.)

## Politics in Middle-earth

Tolkien noted in 1943 that 'My political opinions lean more and
more to Anarchy (philosophically understood, meaning abolition
of control, not whiskered men with bombs) — or to "unconstitu-
tional" Monarchy.' I have already mentioned his hostility to the
state. Actually, whiskered or not, Tolkien arguably anticipated the
eco-sabotage of the group Earth First!; his approval stretched to
the war-time 'dynamiting [of] factories and power-stations; I hope
that, encouraged now as "patriotism," may remain a habit! But it
won't do any good, if it is not universal.'

Some years later, Tolkien wrote:

> I am not a 'socialist' in any sense — being averse to 'planning' (as must be plain) most of all because the 'planners,' when they acquire power, become so bad — but I would not say that we had to suffer the malice of Sharkey and his Ruffians here. Though the spirit of 'Isengard,' if not of Mordor, is of course always cropping up. The present design of destroying Oxford in order to accommodate motor-cars is a case. But our chief adversary is a member of a 'Tory' Government.

(He was referring to a narrowly-defeated proposal in 1956 to put a 'relief road' through Christ Church Meadow — something with a distinctly contemporary ring.)

So Tolkien himself can be classed as an anarchist, libertarian, and/or conservative — not at all in the contemporary sense of the last (which has been almost entirely taken over by neo-liberalism), but in the sense of striving to conserve what is worth saving. None of these categories can easily be assimilated to either Left or Right, which is itself usually sufficient cause to be dismissed by those who like to have these things cut and dried. In a consistently pre-modern way, Tolkien was neither liberal nor socialist, nor even necessarily democrat; but neither is there even a whiff of 'blood and soil' fascism. In this, he contrasts strongly with modernists such as T. S. Eliot, Ezra Pound, D. H. Lawrence and Wyndham Lewis: writers to whose work that of Tolkien is frequently unfavourably compared. But this is no surprise; Tolkien was trying to do something completely different. Consider too that besides imperialistic nationalism, of which Tolkien was very suspicious, something common to all strands of fascism (but especially Nazism) is the worship of technological modernism, which he positively hated.

That antipathy is obvious throughout his works, down to the background detail of, say, the fall of Númenor (Tolkien's Atlantis) through *hubris*, which consisted of both domestic political autocracy, including the suppression of dissent, and a foreign policy based on technological and military supremacy. Actually, German Nazism was a particular tragedy for Tolkien. In 1941, he wrote to his son Michael that 'I have in this War a burning private grudge' against Hitler, for 'ruining, perverting, misapplying, and making for ever accursed, that noble northern spirit, a supreme contribution to Europe, which I have ever loved, and tried to present in its true light.'

It is also noteworthy that when the German publishers of *The Hobbit* wrote to Tolkien in 1938 asking if he was of *'arisch'* (aryan) origins, and could prove it, he refused to do so, indignantly replying that 'if I am to understand that you are enquiring whether I am of *Jewish* origin, I can only reply that I regret that I appear to have *no* ancestors of that gifted people.' He consequently advised Allen & Unwin that he was inclined to 'let a German translation go hang.'

Nor is Middle-earth fascist, let alone Nazi. The Shire, for example, functions by a sort of municipal (not representative) democracy, which Tolkien himself accurately described as 'half republic half aristocracy.' The former half has, typically, been ignored by Tolkien's critics in their eagerness to assail the latter; but even here, their case is mixed at best. On the one hand, there is undeniably a certain amount of quasi-feudal paternalism and deference in the Shire, which is particularly evident, and sometimes annoying, as in the case of Sam. To me (and I doubt I am alone in this), it reads like a relic, and is far too hard to take seriously to offer any kind of model whatsoever. Similarly, of the three positions of authority in the Shire, two are hereditary and only one elected. But these officers' powers, and duties, are minimal. True, by the end of *The Lord of the Rings* there is again a King; and one whose kingly qualities Tolkien goes out of his way to establish. But Aragorn merely grants to the Shire, and other areas, the kind of effective independence they already had. Note too that his accession was only with the approval of the people of the City. In other words, local self-government or 'subsidiarity' obtains: most decisions are taken at the lowest possible level, closest to those who are most affected by them.

The Shire as a yeoman-republic thus has strong links to the tradition of civic republicanism, with its emphasis on a self-governing citizenry and its fear of corruption by clique and commerce. As Donald Davie noticed, the implication of *The Lord of the Rings* points firmly 'towards the conviction that authority in public matters' — as distinct from self-government — '. . . can be and ought to be resisted and refused by anyone who wants to live humanely.' This tradition has pre-modern roots, in Aristotle, Cicero and Machiavelli, but its contemporary relevance is none the less for that; and it reminds us that modern parliamentary liberalism has no franchise on democracy and community, or on solutions to our problems — particularly when it has withered to casting a ballot every four or five years for one of two largely overlapping parties. (I once asked Gregory Bateson which political system he thought was best and most humane; he replied, at least half-seriously, 'An inefficient monarchy.')

The Shire also has clear resonances with other postmodern and ecological values that are returning to the fore as modernity turns sour. In sharp contrast to our possessive individualism, the hobbits are intensely communal — *The Lord of the Rings* rarely follows the story of less than two together — and live in a relatively simple and frugal way. Rediscovering the difference between quality of life and standard of living, something hobbits have never forgotten, is becoming urgent. Collective voluntary simplicity is becoming the only positive alternative to collective immiseration. The Netherlands currently requires an area seventeen times its own size to sustain itself; if everyone on Earth lived like an average Canadian, two more Earths would be required to provide the energy and materials needed. And these two places are not the worst. Yet on the whole, most people are not even happier for it!

Other societies in Middle-earth function differently, although mostly under the aegis of non-autocratic royalty. Each is distinct, even those among humans: the Gondorians, the Riddermark and the Men of Bree are not interchangeable. Even if we want to regard hobbits, elves, dwarves and so on as different human 'races' (which would be a crude simplification), none of them suggests that it is possible or even desirable for people to live as 'fortuitously separated brethren,' with their first loyalties to an abstract 'humanity' over and above their own kind and communities. Perhaps such realism too has offended some defenders of this tattered liberal shibboleth.

On the other hand, *The Lord of the Rings* certainly does hold out the hope that different kinds and communities can respect one another's differences, and live at peace with each other. And none of them resembles Mordor: an utterly authoritarian state, with a slave-based economy featuring industrialized agriculture and intensive industrialism — 'great slave-worked fields away south,' while 'in the northward regions were the mines and forges' — all of which is geared towards military production for the purpose of world-wide domination. And it is noteworthy, recalling the intense cults that surrounded such men as Hitler, Stalin and Mao, even in an officially secular state, that Mordor is also an 'evil theocracy (for Sauron is also the god of his slaves) . . .'

To confuse Sauron with the pre-industrial kingships of Gondor or Rohan would be absurd. As Madawc Williams remarks, 'if one king feels morally bound to respect your existing rights while the other is planning either to enslave you or feed you to his Orcs, you'd have little trouble knowing which side you ought to be on!' Furthermore, what is 'The Scouring of the Shire,' politically speaking, but an account of local resistance to fascist thuggery and forced modernization?

That leaves the 'approval of traditional property settlements.' Well, I doubt if Tolkien's approval could have been taken for granted; it would probably have depended a great deal on what was proposed for the land in question. And as Jonathan Bate points out, redistributing ownership is not going to be much use if the land in question is poisoned beyond use.

As I mentioned earlier, Bate makes another important point: a distinction between love of the land and love of the fatherland. The former, which is clear both in Tolkien's personal life and in his books, involves a fierce attachment to highly specific and local places and things. As such, it offers little foothold to the inflated emotional abstractions that are so essential to fascist nationalism. This is vividly illustrated in Sam's saving realization, when tempted by the Ring of Power, that: 'The one small garden of a free gardener was all his need and due, not a garden swollen to a realm; his own hands to use, not the hands of others to command.'

## *Radical Nostalgia*

Cultural materialism not only seems to produce an inability to read, or to recognize other dimensions than power (narrowly meant) and its effects. Remarkably, even in that realm it falls down. Take nostalgia, for example. Fraser Harrison, whom I quoted earlier, agrees with Raymond Williams that 'nostalgia recognizes no duty to history.' He asks us to recognize, however, that:

> there is another dimension to nostalgia and that it should not be dismissed as simply a self-indulgent, escapist and pernicious failing. Whereas its account of history is patently untrue, and more ideological than it would pretend, it does none the less express a truth of its own, which reflects an authentic and deeply felt emotion. . . . Our addiction to it is surely a symptom of our failure to make a satisfactory mode of life in the present, but perhaps it can also be seen as evidence of our desire to repair and revitalize our broken rela-

tions. The pastoral fantasy nostalgia invented is after all an image of a world in which men and women feel at home with themselves, with each other and with nature, a world in which harmony reigns. It is an ideal . . .

Now Tolkien gives us to understand, as strongly as possible while still writing a story and not a tract, that nostalgia pure-and-simple will not suffice. In Middle-earth, it is the Elves whose nostalgia is the strongest — both in the sense of yearning for the past and attempting to maintain that past now, in places like Lothlórien and Rivendell. But the aristocratic and artistic Elves, despite their valiant resistance, plainly offer no real solution to the central problem of the Ring.

Yet it is also true that his work is suffused with the 'pastoral fantasy' of a better world, equally memory and longing, to which Harrison refers. And far from encouraging a passive retreat from political and social realities, such ideals have real power.

## Activism

To pick a local contemporary example, there are (mainly) young people trying, as I write, to defend the remaining countryside outside Newbury, Berkshire, against yet another destructive, expensive and futile bypass. Their principal means of resistance is to put themselves, with extraordinary skill, determination and valour as well as humour, up trees, underground and literally in the way of an army of security guards, bailiffs, contractors and police, not to mention bulldozers and chainsaws. And among them, I found only one person out of dozens who hadn't just read *The Lord of the Rings* but knew it, so to speak, inside out. (Indeed, one of its leaders, if that is the word I want, is one Balin.) It is no coincidence, then, that an early supporter of one such bypass, running through Dartmoor, slammed his opponents as 'Middle Earth Hobbits'! Nobody can tell me that Tolkien's books do not encourage such ecological activism; nor, for that matter, that he himself would not have been firmly on the side of the trees and their protectors.

This is not the only example; once having seen through the lie that Tolkien's books are a bucolic retreat from 'reality' that induces an apolitical passivity or right-wing quietism, others quickly appeared. Like Meredith Veldman, I too found significant common ground between the work of the left-wing historian and peace activist E. P. Thompson and that of Tolkien. What Veldman calls 'the romantic protest movement' unites the CND/END campaign of resistance to nuclear weapons, the ecology movement beginning in the 1970s, and 'Middle-earth as moral protest.' Thus, the countercultural success of Tolkien's book among 60s radicals and dissidents was no anomaly; far from it.

In 1972, David Taggart sailed into a French nuclear testing area, an action which led directly to the founding of Greenpeace. His journal records that 'I had been reading *The Lord of the Rings*. I could not avoid thinking of parallels between our own little fellowship and the long journey of the Hobbits into the volcano-haunted land of Mordor . . .' Nor had it escaped Taggart's notice, or that of other attentive readers, that Mordor's landscape is one of industrial desolation, polluted beyond renewal; and that such desecration is inseparable from its autocratic, unaccountable and unrestrained exercise of political power.

As it happens, this all has particular resonance for me. In addition to my having discovered Tolkien in 1967, the late E. P. Thompson has been one of my mentors, both personally and professionally; and I was an active Greenpeace supporter from the mid-1980s. I still see no serious contradictions here, and no doubt I am not unusual in these combinations. But there is no autobiographical element to my final and most recent example. Here is Maria Kamenkovich on Tolkien in the former USSR, where *The Lord of the Rings* circulated in *samizdat* form: 'Western readers must understand that for us Tolkien was never any kind of "escape." When hobbits laughed at the absurd "distribution," we didn't laugh at all, because the same thing caused millions of deaths among the peasants in the USSR in the 1920s. When Aragorn held up the elf-stone at the parting with the hobbits, we felt desperate because we did not have any hope of winning our battle at home . . .'

Thus the Siege of the White House in Moscow found itself intertwined with the Battle of the Green Fields in the Shire:

> Western friends of Russia know what happened in Moscow on 19–22 of August 1991, but I doubt that they were informed that many people remembered Tolkien when they made barricades from trolley-buses (just like hobbits from country wains!). It is important to note that the first [complete] translation officially published went on sale only a few days before. Moscow members of the Tolkien Society spent all those fearful thunderstorm and rainy nights near the White House holding a defence. The war machines got as crazy as Oliphaunts and stamped down three young archers. And Gandalf stood before the King of Angmar saying: 'You shall not pass . . .'
>
> Tolkien never meant to describe any real events either in the past or the future. But he certainly *added* something to earthly events. It just cannot be helped.

Maybe the chief political problem is not too much fantasy, but *not enough* of the right kind.

## *'Escapism'*

This is a charge which recurs throughout the attacks on Tolkien, and he was familiar with it early in his career. His essay 'On Fairy-Stories' provides his own best defence:

> In using Escape in this way the critics . . . are confusing, not always by sincere error, the Escape of the Prisoner with the Flight of the Deserter. Just so a Party-spokesman might have labelled departure from the misery of the Führer's or any

other Reich and even criticism of it as treachery. In the same way these critics, to make confusion worse, and so to bring into contempt their opponents, stick their label of scorn not only on to Desertion, but on to real Escape, and what are often its companions, Disgust, Anger, Condemnation, and Revolt.

As an instance, he mentions the recent technological innovation (in his time) of mass-produced electric street-lamps. Any writer who ignores such developments, or prefers to discuss, say, lightning, is liable to be labelled escapist: 'out comes the big stick: "Electric lamps have come to stay," they say. . . . [Or:] "The march of Science, its tempo quickened by the needs of war, goes inexorably on . . . making some things obsolete, and foreshadowing new developments in the utilization of electricity": an advertisement. This says the same thing only more menacingly.'

The prison, to encapsulate my theme, is enforced modernity, whose human casualties alone now number in many millions, while for animals and the natural world the holocaust is still continuing. And among its intellectual and cultural warders are the 'realists' and 'rationalists' whom Tolkien has in mind when he says, for example, that 'The notion that motor-cars are more "alive" than, say, centaurs or dragons is curious; that they are more "real" than, say, horses is pathetically absurd.' In the years before Nazism and Stalinism provided such grim confirmation, and before global consumer capitalism took over the job, Tolkien already saw this clearly. Yet his only honour among the élites is still to be accounted escapist, juvenile and irrational. This, after Tolkien, I utterly deny: 'it is after all possible for a rational man, after reflection (quite unconnected with fairy-story or romance), to arrive at the condemnation, implicit at least in the mere silence of "escapist" literature, of progressive things like factories, or the machine-guns and bombs that appear to be their most natural and inevitable, dare we say "inexorable," products.'

*The Lord of the Rings* is hardly escapist within its own context, either, centred as it is around a war, struggle, hardship and suffering. And at the end of his tale, occasional hints about other worlds notwithstanding, Tolkien returns us firmly to *this* one: at the Grey Havens, after the departure of Frodo and Gandalf, Sam 'stood far into the night, hearing only the sigh and murmur of the waves on the shores of Middle-earth, and the sound of them sank deep into his heart.' We stand with him, and his words, the last words of *The Lord of the Rings* — 'Well, I'm back' — apply to the reader as much as to him. The nostalgia Tolkien engenders, therefore, is finally redirected back into our own lives, here and now. In the poet Geoffrey Grigson's still more compact words:

> *be comforted.*
> *Content I did not say.*

# · 3 ·

## MIDDLE-EARTH:
## NATURE AND ECOLOGY

*Last night I dreamed I saw*
*the planet flicker.*
*Great forests fell like buffalo.*
*Everything got sicker.*
*And to the bitter end*
*big business bickered . . .*

'BEYOND COMMUNITY, but not far beyond it, there is nature.'
And there is much more to *The Lord of the Rings* than the Shire and
its hobbits. As Gildor pointed out to Frodo, the wide world cannot be fenced out; and as Merry admitted, there are higher and
deeper things, without which no hobbit could tend his garden in
peace. The Shire is surrounded and supported by the vastness
of Middle-earth, which preceded and will follow it in time as it
encloses it in space. Tolkien himself remarked that 'I have, I suppose, constructed an imaginary *time*, but kept my feet on my own
mother-earth for *place*. . . . The theatre of my tale is this earth, the
one in which we now live, but the historical period is imaginary.'
  'Middle-earth' itself is a modernization 'of an old word for the
inhabited world of Men, the *oikoumene*: middle because thought of

vaguely as set amidst the encircling Seas and (in the northern imagination) between the ice of the North and the fire of the South . . .' (As Tolkien added, it is definitely not, as many early reviewers seemed to assume, another planet!)

## Place

What is most striking about this larger world? Certainly its variety, richness and consistency are extraordinary. The resulting sense of place gives rise to a startling sensation of primary reality. The fact is that Middle-earth is more real to me than many 'actual' places; and if I should suddenly find myself there (which would of course astound me — but not utterly) I would have a better feeling for it, and a better idea of how to find my way about, than if I had been dropped in, say, central Asia or South America. Many others have felt the same way. 'Tolkien's readers all have the same impression: they have walked or ridden every inch of Middle-earth in all its weathers.'

Part of the reason is the remarkable detail and internal consistency of Tolkien's literary craftsmanship. As Barbara Strachey found when she compiled her excellent atlas of Middle-earth, even the phases of the Moon as occasionally mentioned tallied with the dates as given. But more importantly still, it seems that Tolkien was able to imagine places so vividly that he had a profound sense of them to communicate to the reader. There were strong biographical reasons for this. Tolkien was born in South Africa, but after the death of his father he and his mother returned to England, and between the ages of four and eight (also the last four years of the nineteenth century) he grew up in Sarehole, then a small village outside Birmingham, in the West Midlands. He later recalled that 'It was a kind of lost paradise.' He had returned to England 'with a memory of something different — hot, dry and barren — and it intensified my love of my own countryside. I could draw you a map of every inch of it. I loved it with an intensity of love that was a kind of nostalgia reversed. It was a kind of double coming home . . .'

The sense of place comes through powerfully in Tolkien's fiction, and it is reinforced by his naming of places, which also reflects his love and knowledge of language. As a result, there is none of the sense of arbitrariness which attends most invented words (or worlds), but rather one of historical depth and integrity. This localness marks the difference between a conservationist love of the land and a conservative (or worse) love of an inflated fatherland. The ancient reputation of the Swiss for both steadfast republican liberty and an emotional attachment to their unique locale is no coincidence, and may well remind us of the Shire. In terms of my schema in this book, it is the point at which the Shire and Middle-earth coincide.

## Nature in Middle-earth

But Middle-earth far exceeds the Shire, and what is most striking about it is the profound presence of the natural world: geography and geology, ecologies, flora and fauna, the seasons, weather, the night-sky, the stars and the Moon. The experience of these phenomena as comprising a living and meaningful cosmos saturates his entire story. It wouldn't be stretching a point to say that Middle-earth itself appears as a character in its own right. And the living personality and agency of this character are none the less for being non-human; in fact, that is just what allows for a sense of ancient myth, with its feeling of a time when the Earth itself was alive. It whispers: perhaps it could be again; perhaps, indeed, it still is. And there is an accompanying sense of relief: here, at least, a reader may take refuge from a world where, as in a hall of mirrors gone mad, humanity has swollen to become everything, and the measure of everything. Escaping a bloated solipsism, there is a sense of perspective, context, and sanity.

Even the various races of people in Middle-earth are rooted to and unimaginable — both to themselves and to us — without their natural contexts. As Sam said of the Elves in Lothlórien, 'Whether they've made the land, or the land's made them, it's hard to say . . .' Only Gandalf and, increasingly, Frodo, are wanderers with no true home in Middle-earth. The Elves are a special case, having as it were one foot in the glades of Middle-earth and one foot in their ancient home over the Sea; but that very division had tragic consequences for them, in the various conflicts it caused.

Animals appear in Middle-earth and, as Paul Kocher notes, 'Tolkien is sure that modern man's belief that he is the only intelligent species on Earth has not been good for him' (to say nothing of its effect on animals). But he obviously had a particular affection for flora. I count sixty-four species of non-cultivated plants specifically mentioned in *The Hobbit* and *The Lord of the Rings* — surely an unusual number for any work of fiction — in addition to his own invented (or, as it were, discovered) kinds: *athelas, mallorn, lebethron, elanor, niphredil, simbelmynë,* the pale white flowers of Morgul Vale, and the Huorns of Fangorn. Pride of place, however, goes to trees.

## Forests, Woods and Trees

Every forest in Middle-earth — Mirkwood, the Old Forest, Fangorn, even Woody End in the Shire — has its own unique personality. And none is more memorable than the green city of Caras Galadhon in Lothlórien, 'the heart of Elvendom on earth,' the height of whose mallorn-trees 'could not be guessed, but they stood up in the twilight like living towers. In their many-tiered branches and amid their ever-moving leaves countless lights were gleaming, green and gold and silver.'

Incidentally, these colours receive repeated emphasis. Treebeard's two drinking vessels glow, 'one with a golden and the other with a rich green light; and the blending of the two lights lit the bay, as if the sun of summer was shining through a roof of young leaves.' The light in Sam's mind, trapped in the darkness with Shelob, 'became colour: green gold, silver, white;' and when he awoke in Ithilien, through the leaves of the beech-trees overhead 'sunlight glimmered, green and gold.' Even Théoden's bier was green and white, 'but upon the king was laid the great cloth of gold . . .' The traditional association of gold and silver with the Sun and Moon is plain enough, but as anciently valued I think they also symbolize human civilization, by whose reiterated contiguity with green Tolkien meant to convey a harmonious relationship between humankind and nature. Indeed, an inseparable relationship: when asked rhetorically, 'Do we walk in legends or on the green earth in the daylight?,' Aragorn rightly replies that 'A man may do both. . . . The green earth, say you? That is a mighty matter of legend, though you tread it under the light of day!'

Tolkien does not romanticize nature, however. You can easily freeze to death, die of overexposure, drown or starve in Middle-earth. Consider these remarks by Angela Carter on the wood in Shakespeare's *A Midsummer Night's Dream:*

> The English wood is nothing like the dark, necromantic forest in which the Northern European imagination begins and ends, where its dead and the witches live. . . . For example an English wood, however marvellous, however metamorphic, cannot, by definition, be trackless. . . . But to be lost in the forest is to be lost to *this* world, to be abandoned by the light, to lose yourself utterly with no guarantee you will either find yourself or else be found, to be committed against your will — or worse, of your own desire — to a perpetual absence from humanity, an existential catastrophe. . . . The wood we have just described is that of nineteenth-century nostalgia, which disinfected the wood, cleansing it of the grave, hideous and elemental beings with which the superstition of an earlier age had filled it. Or rather, denaturing, castrating those beings until they came to look like those photographs of fairy folk that so enraptured Conan Doyle.

The interest of this passage for us lies mainly in how it *doesn't* apply to Middle-earth. In fact, such 'denaturing,' which transformed Tolkien's beloved Elves from 'a race high and beautiful, the older Children of the world . . . the People of the Great Journey, the People of the Stars' into the wee 'fairy folk' he so hated, was exactly what Tolkien held against Shakespeare. The hobbits may go rambling through an English wood on a day's outing, but as Bilbo soon learned (and as any reader of *The Hobbit* could tell you), wandering off the path in Mirkwood definitely amounted to an 'existential catastrophe.' Tolkien made no attempt to prettify 'the hearts of trees and their thoughts, which were often dark and strange, and filled with a hatred of things that go free upon the earth, gnawing, biting, breaking, hacking, burning: destroyers and usurpers.'

Individual trees figure importantly too. *The Lord of the Rings* begins with the old Party Tree, and ends with a new one. (It nearly ends prematurely with Old Man Willow.) The tree that blossoms in the courtyard in Minas Tirith is a scion of Telperion the White, which with Laurelin the Golden is one of Tolkien's cosmogonic trees of life. In the internal mythology of Middle-earth, they embodied the first light in the universe, and before they died bore a great silver flower and golden fruit: the Sun and Moon. Their light otherwise only remains visible only in the 'star' of Eärendil (that is, Venus). And, of course, hobbits were not Tolkien's only unique creation; he also gave us Ents, and the unforgettable character of Treebeard.

## The War on Trees

When asked the cardinal question in any kind of war — in fact, the question that is itself (however discreet) the first act of war (however polite): 'Whose side are you on?' — Treebeard replies, "'I am not altogether on anybody's *side*, because nobody is altogether on my *side*, if you understand me: nobody cares for the woods as I care for them, not even Elves nowadays. Still, I take more kindly to Elves than to others. . . . And there are some things, of course, whose side I am altogether not on; I am against them altogether: these — *burárum"* (he again made a deep rumble of disgust) "— these Orcs, and their masters".'

Without any suggestion of exact substitution, it is easy to hear the voice of Tolkien himself here. He freely acknowledged his own 'tree-love,' writing that 'I am (obviously) much in love with plants and above all trees, and always have been; and I find human maltreatment of them as hard to bear as some find ill-treatment of animals.' In a letter to *The Daily Telegraph*, published on 4 July 1972, Tolkien objected to an editorial description of Forestry Commission plantations as possessing 'a kind of Tolkien gloom.' Probably writing in view of his 'totem tree,' a Birch in his front yard, he pointed out that:

> In all my works I take the part of trees as against all their enemies. Lothlórien is beautiful because there the trees were loved. . . . It would be unfair to compare the Forestry Commission with Sauron because as you observe it is capable of repentance; but nothing it has done that is stupid compares with the destruction, torture and murder of trees perpetrated by private individuals and minor official bodies. The savage sound of the electric saw is never silent wherever trees are still found growing.

Was Tolkien exaggerating? Let us briefly look at the situation in Britain, which is probably about average in a global context. Half of the remaining ancient woodlands have been destroyed in the last fifty years — as much as in the last four centuries; only 10% of the country is now forested at all, and most of that is non-native coniferous. The Forestry Commission, supposedly owner of woods on behalf of the nation, is being privatized (sold off to the highest bidders for profit) by stealth. Another priority of those in power has been to build, usually through pristine countryside, yet more roads for the unsustainable use of cars. This was never more vividly symbolized than when the 250-year-old Sweet Chestnut tree on St George's Green, in Wanstead, East London, was smashed down and cut up, after determined but non-violent resistance, on 10 December 1993.

Nor is the situation better elsewhere. The magnificent forests of the Pacific North-west, in Oregon, Washington and British Columbia — some groves of trees 1,600 years old, and home to a fantastic array of flora and wildlife — are being felled at a rate that exceeds that of Brazil, leaving clear-cut moonscapes. Roughly an incredible 90% of Western and Central Europe's original temperate forests have already disappeared. In both cases, the remaining pockets (and that is all they are) of old-growth are going fast, along with their lynx, wolf and bear, mostly to be replaced with factory forests: the lifeless 'green deserts' of monoculture timber plantations. Transnational timber companies are hungrily eyeing the last big temperate forests in Eastern Europe and the former USSR. Meanwhile, of primary (untouched) tropical forest — which only occupies 6% of the Earth's land surface, but contains half its species — about 40% has already gone; every year, another area the size of England and Wales is felled. And Third World governments use the continuing irresponsibility of rich countries to continue despoiling their own.

'The Dream of the Rood' is an Anglo-Saxon poem from the tenth century, and one which Tolkien knew well. Its author makes it a glory of the tree that it forms the Cross and bears the body of Christ. But as John Fowles comments, 'it is not Christ who is crucified now; it is the tree itself, and on the bitter gallows of human greed and stupidity.'

## The Tree of Life

Even leaving continuity, renewal and joy offered by 'tree-love' to one side — which cannot be done forever — let us be clear that we are talking about living things which cool and filter the air, absorbing pollutants and noise; regulate and purify rainfall, and retain and enrich the soil; produce oxygen (a mature tree can produce enough to meet the annual requirements of 10 people) and provide shelter and shade as well as aesthetic satisfaction, historical continuity and psychological refreshment; give wildlife somewhere to live; and provide renewable resources of timber, compost, fuel, and medicines.

For these attributes alone, trees are worthy of reverence. But they are also living symbols, spiritually and culturally as well as physically. As Jonathan Bate writes, '"romantic ecology" reverences the green earth because it recognizes that neither physically nor psychologically can we live without green things . . .' And of those green things, trees are the oldest and biggest in the world — the elders of the plant kingdom upon which human beings, along with all other living things, depend utterly. There is no substitute for photosynthesis. As such, they embody (more than just symbolize) both continuity with life in and of the past, in the places and times in which they have slowly grown, and faith in its future, measured in the hundreds, and in some cases, thousands, of years they can live to.

It is thus not surprising that trees have been worshipped as sacred in most cultures and times. The Roman philosopher Seneca (*c.* 5 BC–65 AD) wrote:

> When you find yourself within a grove of exceptionally tall, old trees, whose interlocking boughs mysteriously shut out the view of the sky, the great height of the forest and the secrecy of the place together with a sense of awe before the dense impenetrable shades will awaken in you the belief in a god. And when a grotto has been hewn into the hollowed rock of a mountain, not by human hands but by the powers of nature, and to great depth, it pervades your soul with an awesome sense of the religious.

Nearly two millennia later, Robert Louis Stevenson found that 'it is not so much for its beauty that the forest makes a claim upon men's hearts, as for that subtle something, that quality of the air, that emanation from the old trees, that so wonderfully changes and renews a weary spirit.' Not a few people feel the same way today.

Conversely, there is an awful, sick feeling of wrongness when a big tree falls. As Jay Griffiths writes, 'felled trees lying flat' are like 'the horizontal lines of sadness in the human face, or in the human form knocked flat to the ground. Hope, by contrast, is vertical — in the standing tree, in the standing human figure. The only hope for the trees is that enough people will stand up for them, answering an ancient and universal call . . .'

And it is a universal call, extending from the sublime: the World Tree of Yggdrasil, the Biblical Trees of Life and of Knowledge, the Buddha's Bo Tree — through the tribal-cultural: the English May and Apple-tree, Greek Olive and Myrtle, Celtic Oak and Mistletoe — to the touching if slightly ridiculous: our Christmas Trees, blithely transplanted in space and time from pagan Germany to Victorian London. The stone groves in Gothic cathedrals, honouring long-lost arboreal ancestors, still inspire wonder; and more local comrades, like our long-suffering urban trees, affection. In the words of the historian of comparative religion, Mircea Eliade, 'the tree represents — whether ritually and concretely, or in mythology and cosmology, or simply symbolically — the *living cosmos*, endlessly renewing itself.'

Tolkien would have been particularly aware of Yggdrasil, the World Tree and *axis mundi* (centre of the known world) of the Norse and Germanic worldview, and one which precedes and survives the gods themselves. It was sometimes thought to be an Ash, although the self-renewing and evergreen Yew seems a stronger candidate. Besides these two, other symbolic local trees include the Oak, sacred to Thor, and the Apple, 'the favourite fruit-bearing tree of the people of the North,' its fruit the gift of choice from Nob and Bob to Samwise, with a typically Tolkienian emphasis on the plain people and simple pleasures, upon the Ring's departure from Bree. But what is important here is the mutual dependence of the universal and the particular. To quote Eliade again, 'the Whole exists within each significant fragment . . . because every significant fragment reproduces the Whole.' Thus 'in the dialectic of the sacred, a part (a tree, a plant) has the value of the whole (the cosmos, life), a profane thing becomes a hierophany. Yggdrasil was the symbol of the Universe, but to the Germans of old any Oak (or Ash) tree could become sacred if it partook of the archetypal condition, if it "repeated" Yggdrasil.'

In this way, crucially, a universal sacred symbol is brought back to particular and unique things, places and people (both human and non-human). This is what nourishes the sense of local distinctiveness that is so important for resisting the homogenization of modernity, whereby everywhere and thence everything becomes more or less the same. Such local roots also resist manipulation by abstract political ideologies.

## Tolkien and Trees

Tolkien once referred to *The Lord of the Rings* as 'my own internal Tree.' It was not the only one: 'I have among my "papers",' he once wrote, 'more than one version of a mythical "tree," which crops up regularly at those times when I feel driven to pattern-designing . . . the tree bears besides various shapes of leaves many flowers small and large signifying poems and major legends.' The reference, or application, to his own short story 'Leaf by Niggle' is obvious: Niggle's surviving painting 'Leaf' was but a tiny fragment of the Great Tree of his ambition and final (spiritual) achievement.

Tolkien's trees, whether 'internal' or 'external,' are indeed mythic. In the context of the hallowed place of trees in mythology — of which, as I have said, he was well aware — his dendro-philia was more than a mere personal idiosyncrasy. His 'totem' Birch tree, for example, is sacred to indigenous peoples through-out North America, Europe and Asia. Just these kinds of val-ues, rooted in an enchanted world, are still found among sur-viving indigenous peoples. Their rediscovery, and a consequent re-enchantment, is one of the keys to our collective future sur-vival, let alone renewal; for 'disenchanted' people will fall for the first rationalization for exploiting and destroying, and a disen-chanted world doesn't feel worth defending.

Tolkien's involvement with trees combined the mythically res-onant with the personally poignant in a way which led to an ex-traordinarily vivid depiction in art. He would have liked John Fowles' avowal that 'If I cherish trees beyond all personal (and perhaps rather peculiar) need and liking of them, it is because of this, their natural correspondence with the greener, more myste-rious processes of the mind — and because they also seem to me the best, most revealing messengers to us from all nature, the nearest its heart.'

But Tolkien's trees are never *just* symbols, and in their indi-viduality convey the uniqueness and vulnerability of 'real' trees. One was a 'great-limbed poplar tree' outside his house in the late 1930s, an inspiration for 'Leaf by Niggle,' that was 'suddenly lopped and mutilated by its owner, I do not know why. It is cut down now, a less barbarous punishment for any crimes it may have been accused of, such as being large and alive.' As Kim Taplin remarks, after Tolkien, 'The wanton felling of trees may have allegorical overtones, but it is also an actual evil . . . and the pres-ence of trees, and the love of trees, are actual as well as symbolic goods.'

Tolkien was also historically minded, and his trees have deep historical as well as mythological and psychological roots. Thus, Middle-earth's own Old Forest was not so-called 'without reason, for it was indeed ancient, a survivor of vast forgotten woods . . .'

But even in the Third Age, those were already a thing of the past. And at the opening of the story in *The Lord of the Rings* — itself supposedly in the (imaginary) past of our world — even such remnants are on the edge of doom. On the very border of Fangorn Forest, as Treebeard says, Saruman

> is plotting to become a Power. He has a mind of metal and wheels; and he does not care for growing things, except as far as they serve him for the moment.
>
> . . . Down on the borders they are felling trees — good trees. Some of the trees they just cut down and leave to rot — orc-mischief that; but most are hewn up and carried off to feed the fires of Orthanc. There is always a smoke rising from Isengard these days.
>
> . . . Curse him, root and branch! Many of those trees were my friends, creatures I had known from nut and acorn; many had voices of their own that are lost for ever now. And there are wastes of stump and bramble where once there were singing groves . . .

And if that were not enough, 'it seems that the wind is setting East, and the withering of all woods may be drawing near.' For in what remains of the green garden of Middle-earth, already long tormented by Sauron, has appeared 'the Ring of Power, the foundation of Barad-dûr and the hope of Sauron.' (It is also the hope of Saruman, of course; but he is no more than one of Mordor's imitators and servants.) 'The Ring! What shall we do with the Ring, the least of rings, the trifle that Sauron fancies?' Elrond alone permits himself any irony, even as he too, like all the good and great, acknowledges his helplessness before the Ring on the hand of its maker and master.

## The Ring

Here we must tread carefully, for Tolkien has warned us repeatedly against an allegorical or topical reading of his story, in which

elements receive a more-or-less literal or one-to-one interpretation. His dislike of allegory, as I have already mentioned, is well-
established: 'I think that many confuse "applicability" with "allegory"; but the one resides in the freedom of the reader, and the
other in the purposed domination of the author.' Or as he once
wonderfully complained, 'To ask if the Orcs "are" Communists is
to me as sensible as asking if Communists are Orcs.' And he was
right. The slightly strained quality of that remark stems from his
desire to protect what he had worked so hard to create, namely a
literary artefact that precisely *isn't* 'allegorical or topical.' Without
suggesting that the meaning of the Ring is thereby exhausted,
however, I shall avail myself of my right as a reader to perceive
'applicability' — an application that is becoming ever-increas-
ingly difficult to avoid.

Consider that the Ring epitomizes the strongest economic and
political power in Middle-earth, which already threatens to dominate all others in one vast autocratic realm. There is no greater
power in the material realm. True, it cannot create beauty or understanding or healing, but the three Elven Rings that can 'are ultimately under the control of the One.' And from the point of
view of *those* kinds of considerations, the One Ring's transformative power is unavoidably destructive. Furthermore, this potential
will be realized to the full once the Ring is entirely under the control of Sauron. (Such a potential was prefigured in *The Hobbit*,
when Tolkien observed of the Goblins, Sauron's chief servants,
that 'It is not unlikely that they invented some of the machines
that have since troubled the world, especially the ingenious devices for killing large numbers of people at once, for wheels and
engines and explosions always delighted them . . .')

## *Magic* vs. *Enchantment*

In his essay 'On Fairy-Stories,' Tolkien explicitly links this sort of
scientific-technological ingenuity with *magic*, culminating in 'the

vulgar devices of the laborious, scientific, magician.' The magic of faërie, by contrast, is what he calls *enchantment*. This is a brilliant and vitally important distinction.

Tolkien describes enchantment as 'the primal desire at the heart of Faërie: the realization, independent of the conceiving mind, of imagined wonder.' But the 'realization' here is ambiguous, and properly so; it signifies both the making of the natural world wondrous through the creation of a 'Secondary World . . . artistic in desire and purpose,' *and* the realization (through the former) that the Primary or 'real' world actually is wondrous. In the context of *The Lord of the Rings*, enchantment is the art of the elves; and as such, it has a special affinity with nature both as its principal inspiration and as the object of its enhancement: 'Their "magic" is Art, delivered from many of its human limitations: more effortless, more quick, more complete . . .' and even though the Elves 'became sad, and their art (shall we say) antiquarian . . . they also retained the old motive of their kind, the adornment of earth, and the healing of its hurts.'

Magic too concerns the Earth, but in a completely different way:

> Enchantment produces a Secondary World into which both designer and spectator can enter, to the satisfaction of their senses while they are inside; but in its purity it is artistic in desire and purpose. Magic produces, or pretends to produce, an alteration in the Primary World . . . it is not an art but a technique; its desire is *power* in this world, domination of things and wills.

The Enemy is thus 'Lord of magic and machines,' who favours '"machinery" — with destructive and evil effects — because "magicians," who have become chiefly concerned to use *magia* for their own power, would do so (do do so).'

That power doesn't always start off as pure self-aggrandizement; probably rarely so, in fact. Tolkien recognized that 'frightful evil can and does arise from an apparently good root, the desire to benefit the world and others — speedily and according to the benefactor's own plans . . .' Even Sauron's rise to power at the beginning of the Third Age started 'slowly, beginning with fair motives: the reorganizing and rehabilitation of the ruin of Middle-earth, "neglected by the gods," he becomes a reincarnation of Evil, and a thing lusting for Complete Power . . .' Remember what Saruman the collaborator tried to tempt Gandalf with: 'Knowledge, Rule, Order.'

On historical grounds alone, Tolkien is quite correct; the appropriation of magic and its transformation into modern science is one of the most important events (and closely-guarded secrets) of the past three centuries. And in contemporary terms, the domination of financial and technological magic over enchantment — often through exploiting it (something at which advertising and public relations are masters) — is something we see confirmed everywhere in Middle-earth today, just as we continue to hear a great deal about how all this Progress is not only good for us, but unavoidable in any case. As he wrote in a letter:

> So we come inevitably from Daedalus and Icarus to the Giant Bomber. It is not an advance on wisdom! This terrible truth, glimpsed so long ago by Sam Butler, sticks out so plainly and is so horrifyingly exhibited in our time, with its even worse menace for the future, that it seems almost a world wide mental disease that only a tiny minority perceive it.

Let me be clear: science as a human activity has perfectly honourable antecedents, and is not intrinsically or necessarily perverted by power-as-domination. Even today, some scientists are more oriented to the wonder of the natural world (i.e. enchantment) than its manipulation and exploitation (i.e. magic). Actually, this is discernible within Tolkien's work. In a letter, he observed that

The Elves represent, as it were, the artistic, aesthetic and purely scientific aspects of the Humane nature raised to a higher level than is actually seen in Men. That is: they have a devoted love of the physical world, and a desire to observe and understand it for its own sake and as 'other' . . . not as a material for use or as a power-platform.

The Noldor, or Loremasters, in particular, 'were always on the side of "science and technology," as we should call it . . .' On the other hand, it was the Noldor who cooperated with Sauron in forging the Rings of Power, and were thus duped and betrayed by him.

Nor is technology as such evil, although there is far too much self-interested nonsense about it being 'neutral'; there is nothing morally neutral about a bomb compared, say, with a bicycle. Tolkien admits that 'It would no doubt be possible to defend poor Lotho's introduction of more efficient mills; but not Sharkey and Sandyman's use of them' — and still less, in Treebeard's words, 'orc-work, the wanton hewing . . . without even the bad excuse of feeding the fires . . .' I think the same point is evident from the Dwarves, who were created by Aulë the Smith, and in their hands 'still lives the skill in works of stone that none have surpassed.' They are also constitutionally prone to greed for gold and precious stones, not to mention *mithril*. But when Gimli discovers the Caverns of Helm's Deep, he is adamant that 'No dwarf could be unmoved by such loveliness. None of Durin's race would mine those caves for stones or ore, not if diamonds and gold could be got there. . . . We would tend these glades of flowering stone, not quarry them.'

Nor is science the whole problem, even today. *Nonetheless*, it is true, and vital to admit, that modern science — the ideology of which is sometimes called scientism — is a very different matter. It has become almost inseparable from both power and profit, and sometimes an object of worship in its own right. As such, it is now as much of a problem in our Middle-earth as it is in Tolkien's literary creation.

## The Ring as Megamachine

Virtually every major character in *The Lord of the Rings* refuses to accept the Ring, knowing that no matter how morally strong, they could not resist its power. Significantly, only a hobbit — the member of a humble and provincial, even parochial race, and one close to the Earth — becomes a Ringbearer. But Gollum, originally a hobbit, is more pitiable than evil because he is so palpably its victim — like a tribesman from the Stone Age encountering modern weapons, bulldozers and bureaucracy, who tries (although neither wanted nor needed) to become their servant. In the end, of course, even Frodo fails the ultimate test. And if the Ring is taken, then the Shire will be no refuge. Tom Bombadil alone is completely unaffected by this supreme talisman of power. As Gandalf says, 'the Ring has no power over him. He is his own master. But he cannot alter the Ring itself, nor break its power over others.' Nor does it appear that he alone could withstand the coming of Sauron repossessed of the Ring.

Although not (to my taste) Tolkien's most felicitous character, Tom Bombadil is clearly a *genius loci* who embodies 'the blind grace resident in Nature,' and 'more specifically . . . of the land itself.' He symbolizes, in Tolkien's own words, 'the spirit of the (vanishing) Oxford and Berkshire countryside.' But the point about Bombadil in this context is that, as Galdor says, 'Power to defy our Enemy is not in him, unless such power is in the earth itself. And yet we see that Sauron can torture and destroy the very hills.' That fact becomes brutally clear in Frodo and Sam's agonizing journey to Mordor. It is worth quoting at some length what they found before its door:

> Here nothing lived, not even the leprous growths that feed on rottenness. The gasping pools were choked with ash and crawling muds, sickly white and grey, as if the mountains had vomited the filth of their entrails upon the lands about.

High mounds of crushed and powdered rock, great cones of earth fire-blasted and poison-stained, stood like an obscene graveyard in endless rows, slowly revealed in the reluctant light.

They had come to the desolation that lay before Mordor: the lasting monument to the dark labour of its slaves that should endure when all their purposes were made void; a land defiled, diseased beyond all healing — unless the Great Sea should enter in and wash it with oblivion. 'I feel sick,' said Sam. Frodo did not speak.

Later, entering Morgul Valley, Frodo observed that 'Earth, air and water all seem accursed.' And closer still to Mount Doom, they found 'a huge mass of ash and slag and burned stone,' where 'the air was full of fumes; breathing was painful and difficult . . .' Do we not see such blighted industrial wasteland today in Eastern Europe and Russia, and could we not easily find its equivalents elsewhere in 'the West': radioactive deserts; poisoned rivers and even seas; clearcut and slashed and burned acres that were once rainforest, richest in life anywhere on the planet; smoking, reeking cities where life, by contrast, is cheap? All this has a name, by the way. The Greek *oikos*, which gives us 'eco,' means home or abode; the Latin *caedere*, to kill; hence, ecocide. (And the combination of Greek and Latin only confirms that no good can come of it.)

Tom Shippey has observed that the Ring is (1) immensely powerful, (2) dangerous, even lethal, to all its possessors, and (3) will ultimately triumph if it is not destroyed. Thus 'it is a dull mind which does not reflect, "Power corrupts, and absolute power corrupts absolutely".' And in addition to the distinctively modern nature of this understanding, he also shrewdly reminds us that the Ring is addictive in a way — let's call it 'lifestyle' — that we are all now familiar with. This interpretation can be further tightened up with no loss of meaning, indeed in a way that brings matters sharply home. It needs no allegorical special pleading or stretch of the imagination to see that *our* Ring is the malevolent amalgam of the unaccountable nation-state, capitalism in the form of transnational economic power, and scientism, or the monopoly of knowledge by modern technological science. Like Tolkien's Ring, there are apparently no limits to its potential mastery of nature (certainly not those of Mercy), and, once it is on the finger of its collective principal servants — that is, completely removed from any democratic accountability — no way to control it.

The Ring's servants have no wish to control it, of course; rather, to feed it. Tolkien noted in 1945, 'As the servants of the Machines are becoming a privileged class, the Machines are going to be enormously more powerful. What's their next move?' There is precious little control as things are. Sporadic public protest and non-governmental organizations worry away at its edges and fight 'the long defeat,' as Galadriel called it — but always under the shadow of 'that vast fortress, armoury, prison, furnace of great power, Barad-dûr, the Dark Tower, which suffers no rival, and laughs at flattery, biding its time, secure in its pride and its immeasurable strength.'

This is so not least because, in a twist even Sauron never thought of, most people — even those who are already living in ways that constitute the solution to its terrible problems, and will suffer the most by its adoption — seem so seduced by the mega-machine's handmaidens in advertising, the media and the movies that they can hardly wait to sign up; addictive indeed.

Tolkien has frequently been accused of a simple-minded moral Manicheism, simply pitting good against evil. Whether in relation to individuals (as I have already mentioned) or races, this charge is wide of the mark. One of the glories of Middle-earth is its messy pluralism; the alliance against Mordor is only just cobbled together (thanks mainly to Gandalf) among people with drastically different cultures, languages, habits, and agenda. *The Lord of the*

*Rings* celebrates such difference and pleads, as Shippey says, 'for tolerance across an enormous gap of times and attitudes and ethical styles.'

These are precisely the things that are jeopardized by Sauron, who seeks to turn all Middle-earth into one vast homogeneous entity, under his all-seeing Eye that might remind us not only of 'single vision and Newton's sleep,' in Blake's words, but Foucault's alarm-call about the insidious growth of institutionalized knowledge-as-power: 'Where religions once demanded the sacrifice of bodies, knowledge now calls for experimentation on ourselves, calls us to the sacrifice of the subject of knowledge.'

Thus Edward Teller, 'father' of the hydrogen bomb, speaks for many scientists, and their corporate backers, when he states flatly that 'There is no case where ignorance should be preferred to knowledge . . .' That may well be true for science; it is by no means always true for humanity or the world. Let us recall that Saruman's thirst for knowledge at all costs — of the 'magical,' including scientific, kind — was precisely what baited Sauron's trap in which the wizard was caught. And recalling Tolkien's distinction between magic and enchantment permits us to recognize modern profit-driven and state-protected science for what it is: not the disenchantment (or demystification, or rationality) that they pretend, but modernist magic: a powerful *counter*-enchantment, much of whose power stems from being a spell that denies that it is one: a secular religion, literally a bad faith. With better reason than he knows, Teller's interviewer described him as 'our great master of the black art of detachment.' As Adorno and Horkheimer recognized,

> In the enlightened world, mythology has entered into the profane. In its blank purity, the reality which has been cleansed of demons and their conceptual descendants assumes the numinous character which the ancient world attributed to demons. . . . It is not merely that domination is paid for by the alienation of men from the objects domi-

nated: with the objectification of spirit, the very relations of men — even those of the individual to himself — were bewitched.

Modern magic/science was itself literally born of a dream: that of Descartes, a founding father of modernity (and patron saint of animal vivisection), on the night of 10 November 1610, of 'the unification and illumination of the whole of science, even the whole of knowledge, by one and the same method: the method of *reason.*' This dream eventually combined with the boundless ambitions of Francis Bacon, who advised torturing nature to extract her secrets and further 'the enlarging of the bounds of Human Empire,' boasting of 'leading you to nature with her children to bind her to your service and make her your slave;' and of Galileo, who did so much to further the technique of reducing all merely personal and therefore 'secondary' experience to abstract 'primary' mathematical quantities. As a result, as Horkheimer and Adorno put it, 'What men want to learn from nature is how to use it in order to dominate it and other men. That is their only aim.' (This, of course, is Tolkien's definition of Magic.)

In so doing, they continue, 'The destruction of gods and qualities alike is insisted upon,' along with 'the extirpation of animism.' But note that monotheistic faith collaborates in this program: 'Reason *and* religion deprecate and condemn the principle of magic enchantment.' Neither can long abide anything or anyone escaping the sway of what they need to be total and universal truth; exceptions become anathema.

### *Mordor on Earth*

It has been said, with many variations, that 'Mordor is Wigan or Sheffield,' or Leeds, or Birmingham. But to concentrate too much on Tolkien's anti-industrialism is to miss the larger meaning. Although he did not write the following passage, any reader familiar with Tolkien's work will immediately recognize the terrible authenticity of this description of being inside Mordor:

> Around us, everything is hostile. Above us the malevolent clouds chase each other to separate us from the sun; on all sides the squalor of the toiling steel. . . . And on the scaffolding, on the trains being switched about, on the roads, in the pits, in the offices, men and more men, slaves and masters, the masters slaves themselves. Fear motivates the former, hatred the latter, all other forces are silent. All are enemies or rivals.
>
> . . . This huge entanglement of iron, concrete, mud and smoke is the negation of beauty. . . . Within its bounds not a blade of grass grows, the soil is impregnated with the poisonous seeds of coal and petroleum, and the only things alive are machines and slaves — and the former are more alive than the latter.

This is the essence of Mordor, and although Tolkien wrote *The Lord of the Rings* before the death-camps were widely-known of, he seems to have perceived something essential about the terminus of modernity's merciless logic.

Auschwitz, as is clear from Primo Levi's account, was equally a brutal human or social desecration and a natural or ecological one. The two cannot, in good faith, be separated. Sauron's own campaigns recognize this fact. What first announces his presence, everywhere in Middle-earth, is the ugliness and impoverishment of local natural habitats, 'disgracing the earth,' as William Morris said, 'with filth and squalor.' What follows is the loss of their (remaining) inhabitants' ways of life and independence. And as with us, the first and worst victims are always the weakest and most defenceless: a category that includes trees and animals, as well as children, women, the poor, and the indigenous.

Sauron's strategy is repeated by every avaricious government today: from the wholesale destruction of Tibet, forests and monasteries alike, by China; and Saddam Hussein's campaign against the Marsh Arabs, as much by massive drainage as by weaponry; to Indonesia, where it is accompanied by a smokescreen of 'rehousing' and 'educating' the indigenous people before 'developing' their forest homes, in collaboration with the World Bank. We should also note that the first of these is a communist crime and the last a capitalist. In other words, like the distinction between the destructive exploitation of nature and genocide against humans, this difference too is a secondary one. It is therefore not of much use in getting to grips with the problem which Tolkien addresses.

That is not surprising. Marx had a profound admiration of and respect for capitalism, as is clear in his paeans to its power, and he supported Western imperialism. He had nothing but contempt for tradition and 'rural idiocy,' and along with fetishizing the economic and the 'material' he limited value strictly to whatever had been 'produced' (really, only ever transformed) by human labour; thus there is none whatsoever, according to his system, in nature as such. In the best nineteenth-century way, he also approved of, and wanted to extend, scientistic materialism and rationalism. Lenin changed nothing of all this, admitting only the necessity to give 'historical laws' a helping hand with brutal vanguardism. Even leaving aside its historical track-record, then, exactly what kind of a basis does Marxism still provide for a radical, let alone ecological, alternative to current capitalism? (Democratic socialism, and certain socialist values, are a different matter; and here, a convergence with ecologism may well be possible and desirable.)

Conversely, regeneration of the land strengthens that of its people, and vice-versa. Strong and free societies value and protect their natural contexts (including sacred places), which return that trust by protecting and supporting them. Realizing this and acting on it, there is hope. And Tolkien does suggest that such renewal is possible, as it proved in the Shire after its devastation by Sharkey. But hope demands clear sight of the scale of the problems we face. Taking my cue from Tolkien's work, I have chosen to concentrate here on the decimation of the natural world. But the human cost, both physical and spiritual, is plainly implicit in his chronicle of the Third Age, and it should be understood as an integral part of the whole. Fowles is indulging in no hyperbole when he says that 'In the end what we most defoliate and deprive is ourselves.'

## The War on Life

Of course, such destructiveness is far from new; but it has only become possible in the scale of its recent magnitude and thoroughness in and, more to the point, thanks to modernity. It is in this connection we should recall Tolkien's anti-modernism, which has earned him such a reputation for 'reaction.'

The global ecological holocaust, designed and driven by the Ring of Power, is currently proceeding apace. I have already mentioned deforestation, and the same thing is happening in many other ways. Just in Tolkien's little corner of the Earth, England — and a relatively secure, rich and privileged corner at that — hundreds of thousands of miles of hedgerows have been grubbed up; wetlands are still being drained; on current projections, there will be no remaining lowland peat bogs by 2020, and no estuaries by 2200. Moist lowland hay- and wildflower-meadows are already almost gone, with 95% of them destroyed since 1947.

By 1980, one-quarter of main rivers had been diverted into concrete channels; other waterways have disappeared altogether, due to 'overabstraction' (an apt word for it!) by water companies newly privatized and protected by the government. True, our rivers haven't yet started catching fire, like the pitiable Cuyahoga in Cleveland, Ohio; but 'They're always a-hammering and a-letting out a smoke and a stench, and there isn't no peace even at night in Hobbiton. And they pour out filth a purpose; they've fouled all the lower Water, and it's getting down into Brandywine.' Nor are coastlines safe; huge oil spills seem to have become a regular event. The North Sea itself is a chemical soup, now becoming lifeless. The skylark, bittern and water vole are disappearing, along with the dormouse, turtle dove, tree sparrow, song thrush, linnet, snipe, stone curlew, corncrake, and corn bunting: all victims of intensive farming, drainage and development.

*This picture can be translated worldwide.* Whereas the sad list of endangered, embattled animals seems endless, the number of motor vehicles worldwide (now at about 750 million) increases by 12 million a year. We are running out of water — something like an extraordinary 40% of people (more than two billion) have no reliable clean water and sanitation, while demand is soaring, mainly for intensive agriculture. Our environment is awash with toxic chemicals, whose long-term effects are a complete wild-card, and even the sea — for so long beyond the reach of human effects — is succumbing. Meanwhile the human population could well *double* from its current 5.7 billion by just 2050; the lowest possible projection is a rapid rise to 7.8 billion. We — a single species — are already consuming roughly 40% of the Earth's total biological productivity. For the same reasons, what Richard Leakey and Roger Lewin call 'a biological catastrophe of immense proportions' is now well under way: the mass-extinction by human beings of other species of life, approaching 100,000 a year. At this rate, we shall wipe out half of the Earth's remaining species in the next few decades, including countless species whose roles in 'ecosystem maintenance' we don't even know, and virtually all the larger land animals — elephants, rhinos, tigers, bears, kangaroos and so on (and on). In the process, human beings stand to become their own victims in this ecological holocaust. To quote one botanist, the world is being run as if it was a business in the course of liquidation. But 'If you take a realistic view of the world, we have only got one of them, and it does have to be managed sustainably.'

This débâcle is driven by a combination of the secular gods of production and consumption, the refusal to limit population, and an unquestioned but well-funded scientific technology. The common denominator is a mindless modern humanism, laid bare by

David Ehrenfeld, which has almost nothing in common with the humane and sceptical wisdom of its classical origins. But even some of the technocrats have recognized that all is not well. Do they have an answer? Yes, indeed: more of the same. Thus, the problems resulting from arrogance, meddling and *hubris* will be attacked not with humility, self-restraint, recognizing limits, but with more of the same; not by self-control — anything but that — but by more control. But the latter is, of course, a grand illusion, since each new attempt to control results in unforeseen developments that are perceived in turn as 'out of control' and therefore justifying the next expensive and unsuccessful bout of 'controls.'

Four hundred years ago, before Bacon, Descartes and Galileo had assisted at the birth of the modern revolution, Michel de Montaigne distilled the preceding ages of pagan, classical humanist and Christian thought into his *Essays*. They contain the kind of wisdom that Tolkien's stories — written during modernity's increasingly Pyrrhic and perhaps, literally, final victory — urge us to rediscover. It is something that no amount of mere knowledge can ever replace. Recalling Apollo's command at Delphi to 'Know Thyself,' Montaigne observed that 'Except you alone, O Man, said that god, each creature first studies its own self, and, according to its needs, has limits to its labours and desires. No one is as empty and needy as you, who embrace the universe: you are the seeker with no knowledge, the judge with no jurisdiction and, when all is done, the jester of the farce.'

## Selling Ourselves

Today, much of what Tolkien called 'the utter folly of these lunatic physicists' has captured the life sciences. Perhaps he wouldn't have been surprised. After all, having already deliberately bred orcs (the Uruk-hai) and trolls that do not weaken in sunlight, Saruman and Sauron were already well on their way to genetic engineering. Here, too, biotechnologists and genetic engineers are already patenting, reproducing and selling, at immense profit, cells, genetic material, and now entire species. The clever primate ('Look what I can do!') manipulates DNA, inserting genes from one species into another, engineering animals to develop cancers, releasing transgenic organisms into the wild, producing mice with human ears growing on their backs, pigs with human genes, giant calves. Obsessed by the prospect of profits, he has no idea of the risks; nor, frankly, does he much care.

The tiger is disappearing? Never mind, we'll preserve its frozen DNA. Later, we can see if we can reconstitute a wild animal without any habitat or ecosystem, learning, socialization or natural prey. Genetic samples from endangered indigenous people have already been collected and stored for commercial use. (Sorry, we can't help you with the ethnocide). After all, there are billions of dollars at stake in all this. Meanwhile, is there injustice and ever-increasing inequality of all sorts? Do a few hundred billionaires possess wealth equal to that of billions of the poor? OK, let's make the market global, and put a price on literally everything. And if that's not a kind of universal religion, what is? Welcome to Life, plc.

This sort of report is notoriously numbing, but I make no apology for it. This is something very like what the remaining 'free peoples' at the end of the Third Age in Middle-earth were up against, just as we are at the beginning of the twenty-first century. I don't say there have been no battles won, nor that there are no reasons to hope. There are — indeed, *The Lord of the Rings* is one — and as the overall picture gets darker, the signs and seeds of hope become just that much more important to appreciate and nourish. But as Elrond says, there have been 'many defeats, and many fruitless victories.' Orwell was right, and still is: 'The actual outlook is very dark, and any serious thought should start out from that fact.'

Since hobbits 'have a passion for mushrooms, surpassing even the greediest likings of Big People,' I will summarize the situation in a tiny but telling way. It seems that wild mushrooms are gradually dying out across Europe. At least seventy species of fungi are now extinct; another 600 far less common, including the giant chanterelle, penny-bun or cep, and wood blewits. In Holland, more than 180 species are on the verge of extinction; in Germany the number of chanterelles has dropped from several thousand to a few hundred, while in Britain the once common cep can now increasingly only be found in remote parts. The cause is habitat-loss, combined with acidification from increased levels of nitrogen and sulphur in the air, and heavy metals in the soil. The species most affected are those associated with the roots of trees and other plants; one ecologist has said 'mass extinctions' are now imminent, and fears the consequences for the fungi's symbiotes.

## On 'Sentimentality'

The threat to wild fungi should shake anyone who knows and is fond of hobbits in a way that is quite distinct from an ecologist's apprehension; but it is no more or less valid. Yet if the New World Order can apparently dispense with nature's vital material attributes, what hope is there for moral or aesthetic or cultural considerations? As Richard Mabey writes, these 'are now seen as, at their best, sentimental and impractical, and at their worst — it is a favourite phrase — "purely subjective preferences." Somewhere along the line many deep and widely shared human feelings — an affection for native landscapes, a basic sympathy towards other living things, a feeling of respect for our rural history — have become regarded as a devalued currency.' Or as Fraser Harrison puts it:

> throughout these years, nature has nevertheless prevailed as the richest source of metaphor concerning the human condition. And yet at the time when we have most need of its metaphorical resources, we suddenly discover that nature is

failing us, or rather we are failing nature by destroying it. It is in this sense that I believe we can claim to have our own indispensable *cultural* need of conservation. . . . Apart from all other consequences, the loss of each species or habitat from the countryside amounts to a blow struck at our own identity.

For this reason, the pioneering organization Common Ground has been recently putting 'the cultural arguments about trees and woods alongside the scientific, economic and ecological ones,' pointing out that we need trees for their 'longevity, their beauty, their generosity,' and much else. Tolkien understood this in his bones, and, in his inseparable mix of the natural world and the historico-cultural, conveys it as well as anyone. He would have agreed with Ruskin that

> No air is sweet that is silent; it is only sweet when full of low currents of under sound — triplets of birds, and murmur and chirp of insects, and deep-toned words of men, and wayward trebles of childhood. As the art of life is learned, it will be found at last that all lovely things are also necessary: the wild flower by the wayside, as well as the tended corn; and the wild birds and creatures of the forest, as well as the tended cattle; because man doth not live by bread alone . . .

Yet such a position is frequently the target of near-universal cynicism from academics and the left, as well as from the representatives of old-fashioned material self-interest: a revealing alliance. Indeed, the criticisms are often exactly the same, and found in the same pages, as those directed against Tolkien's work.

Even the normally restrained Keith Thomas, in his widely-acclaimed *Man and the Natural World*, described 'the cult of the countryside' beginning in the eighteenth century as 'in many ways a mystification and an evasion of reality. . . . Encouraged by the ease of travel and by immunity from direct involvement in agricultural processes,' it was 'the educated middle classes' who led the way. 'The irony,' notes Thomas, 'was that the educated tastes of the aesthetes had themselves been paid for by the developments which they affected to deplore.' Thus muddle and inconsistency apparently completely invalidates any values arising in their midst. But in that case, who would 'scape whipping?

Ludmilla Jordanova accuses Thomas of not going far enough. 'Western capitalist society,' she writes, 'sentimentalizes animals and plants while systematically destroying them,' with the middle class

> producing 'culture,' the purpose of which was precisely not to reflect accurately their material environment, but to serve their own interests. . . . Animals can remain commodities, not despite our sentimentalizing them, but because we do so. . . . 'Man' never left centre stage, nature has never been, and will never be, recognized as autonomous.

A gloomy outlook indeed! But is it true? To begin with, cultural conservationists are not necessarily cultural conservatives (in the pejorative sense). Tolkien's position, for example, has acquired a new and distinctly radical meaning — or at the very least, potential meaning — as the crisis which partly motivated its writing has deepened and widened. Secondly, a little humility seems in order. If 'has never been' is already debatable — what kind and amount of evidence, for example, might suffice to back up such a statement? — how much more so is 'will never be'? Can even a professor really comfortably speak of and for 'reality' in this way? And hasn't the overall social 'reality' been one of all-too-human inconsistency and confusion, as well as (rather than simply) unadulterated hypocrisy?

## Life's a Beech

It is also extraordinarily high-handed to dismiss all outrage at the desecration of nature by those of middle-class provenance as necessarily 'affected.' Personally, I never feel so sane and at peace as when I am in the company of broad-leaved trees, the taller and older the better. The most glorious sight I know is sunlight playing in a floating world of green, each leaf perfectly individual yet intricately related to its immediate neighbours, others of the tree and those of the whole forest. They turn the sun back into a young god with a glinting halo, whose filtered light would defy any artist in its ceaselessly intricate and moving patterns, as would the colours. Trees in leaf don't occupy space so much as create it, defining near and far and restoring people (in a way human objects like skyscrapers cannot do) to their proper scale in the scheme of things: small and passing creatures. They breathe, producing that wonderfully cool air that is itself a tonic. They heal and restore. (And if one is really lucky, the only sounds will be their rustling, insects and bird-calls, blessedly free of the roaring, snarling, whining of the machine — the audible dementia of Progress triumphant, which is rapidly becoming inescapable anywhere in the south of England, where I happen to live.)

Conversely, while treelessness is not necessarily hell, hell for me is certainly treeless, like the desert we are turning the world into — no shade, no sanctuary, no mercy. According to some experts on global warming, the climate of the south of England (which some of us actually quite liked as it is, or was) is already on its way to becoming Mediterranean; the south of France will take on that of Spain; Spain, Africa's; and Africa? Surely hell itself, with the sun now its sole, jealous and pitiless God, shrivelling all else.

Now there is no doubt that what I experience in woods is the same 'sanity and sanctity' that exists in the work of Tolkien, another tree-lover and sub-creator of just such trees. I can well imagine what the secular clerisy would have to say about such reflections, of course. Despite my efforts in a 'green' direction, there certainly remain some contradictions between my lifestyle and one that, writ large, would sustain more trees. But I deny anyone's right or ability to deny those efforts and their partial success, or to dismiss my feelings about trees as mere delusion, affectation or 'false consciousness' — on the basis of my class, race, gender or anything else. The value that Tolkien found in trees and put into his books, which I and many of his many readers share, may be many things; but no one can say that they are not 'really' ours, or true, implying that we are under some kind of spell, whereas they (magically enough) are not.

Most recently, an army of university postmodernists, cultural studies experts, philosophers and others have created a new academic industry. In effect, they argue that there is no such thing as nature; that nature is entirely the 'creation' or 'product' of human beings. But this absurdly one-sided credo is simply the intellectual elaboration of what powerful economic and political interests already hold, and are acting on. Humans (meaning themselves) have permanent human possession of 'centre stage,' and are therefore free to do whatever they want with a lifeless, passive set of 'resources.'

This is no armchair struggle. In the battle to determine the fate of irreplaceable primary old-growth forests in North America — not in some impoverished backwater, which can argue that it has no alternative to logging, but two of the richest countries in the world — the contestants are increasingly polarized between 'humanists' (in this case the logging industry and its supporters) and 'deep ecologists' (often under the aegis of the organization 'Earth First!'). In the United States, the former have now organized themselves into a powerful pressure-group, allying aggrieved ranchers, miners, loggers, developers, off-road-vehicle users, corporate front groups and right-wing politicans, called 'Wise Use.' A British equivalent now also exists: 'The Countryside Movement,' which promotes the interest of already unaccountable land-owners, agribusiness and blood-sport enthusiasts. Both organizations feed on resentment — the kind that someone who has been getting away with murder for years feels at finally being identified and censured by a weaker party. (Note too the Orwellian twist in their names.)

For such people, as Robert Pogue Harrison elegantly puts it,

> there can be no question of the forest as a consecrated place of oracular disclosures; as a place of strange or enchanting or monstrous epiphanies; as the imaginary site of lyric nostalgias and erotic errancy; as a natural sanctuary where wild animals may dwell in security far from the havoc of humanity going about the business of looking after its 'interests.' There can be only the claims of human mastery and possession of nature — the reduction of forests to utility. . . . The war is between two fundamentally opposed concepts of the forest. One is the concept of the forest as resource; the other of the forest as sanctuary. . . . [The latter] are forced, however, to speak the language of those whom they oppose. This is precisely the language of usefulness. . . . They must contrive a thousand convincing reasons or unconvincing arguments in favour of the utility of forest conservation. For the moment this is the only language that has a right to speak, for it speaks of the 'rights,' that is to say the economic interests, of humanity. It remains to be seen whether one day a less compromised, less ironic language will become possible — a language of other rights and interests, a language, in short, of other worlds.

John Fowles puts it more bluntly: 'We shall never fully understand nature (or ourselves), and certainly never respect it, until we dissociate the wild from the notion of usability — however innocent or harmless the use.'

*The Lord of the Rings*, on balance, says the same thing. It is true that there remain traces of 'stewardship' over nature. Thus, while people certainly can intervene to assist nature's own ability to flourish, Tolkien sometimes implies that that is required for renewal — a very different proposition. I am thinking chiefly of Sam's gift of seeds from Galadriel, with which he re-treed the

Shire, and Tolkien's suggestion that Middle-earth's beautiful places are that way (solely? surely not) because they were loved. On the other hand — and this is the dominant note — Goldberry tells the hobbits, who have asked if the Old Forest belongs to Tom Bombadil, 'No indeed! . . . The trees and the grasses and all things growing or living in the land belong each to themselves.' And in Lórien, Frodo experiences the truth of this, in a tree: 'He felt a delight in wood and the touch of it, neither as forester nor as carpenter; it was the delight of the living tree itself.'

Compare this with Baja, a motherless toddler from the endangered Aka Pygmies in Central Africa: 'As he came to a towering, smooth tree, he placed his hands against the trunk to steady himself, drew back his head, and stared up at the tree, all the way up to the leafy kingdom of its crown spread out against the sky. He stood that way for ten minutes, now and again gently patting the tree.'

But as William Blake observed sadly in 1799, 'The tree which moves some to tears of joy is in the Eyes of others only a Green thing which stands in the way.' The popular American radio-broadcaster Rush Limbaugh, for example: 'One of the most beautiful sounds you can hear in the world today is a tree being chopped down. The most beautiful thing about a tree is what you can do with it after you chop it down.' Some of the good people of the northwestern United States hold that: 'We should get the old growth cut off . . . it's using up land that could be growing a productive crop.' Meanwhile, in university departments there and the world over, students continue to be trained in industrial forestry. They emerge enabled to see bark, limbs, tops and leaves as 'waste,' trees which have fallen over as 'debris,' old trees as 'decadent' (!), and the insects which live in trees as 'pests.' (Of course, this relentless logic is extended to still closer relatives: 'The breeding sow,' writes a meat company executive in the *National Hog Farmer*, 'should be thought of, and treated as, a valuable piece of machinery whose function is to pump out baby pigs like a sausage machine.')

Baja is far from unusual among indigenous or 'Fourth World' people, for most of whom the Earth and its creatures are sacred, and when used, treated with respect and restraint. Yet before all others, these are the people who are threatened with global genocide:

> The eradication of 'primitive' indigenous peoples who live there is always the first step to converting, say, rainforests into toilet-seats, chopsticks, tissues and disposable plywood moulds for concrete. The fact that these forests are home not only to people who know how to live in them sustainably, but to the most intricate, varied and wondrous life-forms on this planet counts, counts as nothing in the logic of 'economic development.'

This is not something confined to exotic and far-away places. From Irian Jaya to Twyford Down, from Lagos to Vancouver, the same unequal struggle is being fought. As Gimli observed, 'Strange are the ways of Men, Legolas! Here they have one of the marvels of the Northern World, and what do they say of it? Caves, they say! Caves! Holes to fly to in time of war, to store fodder in! . . . None of Durin's folk would mine those caves for stones or ore, not if diamonds and gold could be got there. Do you cut down groves of blossoming trees in the spring-time for firewood?' Sadly, tragically, we do.

## Save Us from the Experts

Such a plea — love and respect for the natural world for its own sake — is a fragile but potentially powerful one, in the way of a new shoot of green growth amid cracks in the concrete. Given more dominant and immediate outlooks, of course, those advocating (like Tolkien) or seeking (like a great many of his readers) 'a new communion with nature' are easily accused of 'indulging in fatuous romanticism,' a mere childish and sentimental fantasy, in the worst sense of the word. And literary experts and professionals look down on people who read Tolkien in exactly the same way.

Such people have no need to feel ashamed, however; far from it. It was not they, for example, but official British 'experts' and 'realists,' who decided to feed vegetarian ruminants on animal flesh and douse them with toxic organo-phosphates, only to be astounded when BSE or 'mad cow disease' resulted. To take another example, in the words of American forester Gordon Robinson, 'you don't have to be a professional forester to recognize bad forestry any more than you have to be a doctor to recognize ill health. If logging looks bad, it is bad. If a forest appears to be mismanaged, it is mismanaged.' And it takes some serious training to be able to ignore the clear and obvious distress and pain of animals you are torturing for trivial or highly questionable reasons in the name of medical science, let alone cosmetics.

Take what abuts the garden of the present Prime Minister of Great Britain (as I write): a 250-acre field, stretching into the distance, and nothing else but electricity poles and a few exposed cottages. It is devoid of life, except for the uniform annual crop. Yet within living memory, there were allotments for local villagers, copses of trees, rose hedges, traditional grassland, all manner of birdlife, ponds with fish, newts and waterfowl. From Essex to Yorkshire, now more typical than not of English agriculture, this is the result of 'chemicals, machines, and the search for profit' (albeit massively subsidized by taxpayers). The same thing is happening almost everywhere, only varying in pace and degree: barbed wire but no orchards; commuters' cars but no buses; wild plants, birds and mammals decimated; privatization and development, while commons and greens disappear; and an enduring legacy of poisons in the soil. Is it really mere 'nostalgia' to point out gross unsustainability (to put it nicely), and ask whether this is really what we want? In even the slightly longer run, who are really the realists here: the official experts, or the amateurs who find such developments appalling and immoral?

It seems to take one of the latter to realize that when it comes to the remaining English countryside, 'We are not talking here about "nostalgia for a non-existent golden age." We are talking about something that is green and living and incomparably beautiful.' To take that on board doesn't rule out human activities, including commerce; but it is a prerequisite for undertaking them in a responsible and sustainable way. It is fitting that Tolkien's own beloved mill and its green environs in Sarehole, where he grew up — and which he was convinced would go under — was saved from the local council and developers by a combination of locals and middle-class incomers: ordinary enough people, all.

As Richard Mabey writes, 'the widespread modern yearning for a relationship with nature and the land based, not on ownership or labour, but on simple delight and sensual and spiritual renewal, is an authentic search for a modern role for the countryside.' Even in huge cities like London, remaining green spaces and 'wild bits' are used, and fellow non-human creatures appreciated, by all classes and sections of the community, who value them as a non-commercialized world, at once natural and communal, to experience and enjoy. And this is just a small but vital part of a deeper truth, what Hazlitt meant by saying that 'Nature is a kind of universal home.' We all participate in nature, so the opportunities it offers to realize, accept (where necessary) and enjoy (where possible) that participation are profoundly democratic.

It must be faced that these are minority voices, and still relatively powerless. Compared to the purely scientific case for conservation — which, because of its humanist assumptions, is fatally vulnerable to narrow human self-interest — the cultural/moral/spiritual one is still rare. Yet as Fraser Harrison writes, 'what must be conserved before anything else is the desire in ourselves for Home — for harmony, peace and love, for growth in nature and in our imaginative powers — because unless we keep this alive, we shall lose everything.'

This hole in the conservation case is, he points out, 'a moral rather than an intellectual failing. All of us, not just conservationists, are suffering from a lack of values: we face a moral, no less than an ecological crisis.' And the survival of anything worth the name 'nature' — and therefore of whatever it means to be human *in relation to* nature — looks increasingly likely to depend on the success of just such a case. With the entry of this dimension, however, we are at the very edge of Middle-earth. Although we are still in Tolkien's world, we have been brought up short by the Sea.

· 4 ·

# THE SEA: SPIRITUALITY
# AND ETHICS

*Human existence so embedded in the
bosom of Nature is possible only when and
where Nature is imbued with the life of
divine spirits.*

THE SHORE OF MIDDLE-EARTH marks the literal and symbolic limit of both the natural world, itself enfolding the Shire,
and thus also the domination, actual or potential, of the Ring. But
there is a limit to both, as Legolas said when he recalled first hearing the gulls at Pelargir: 'I stood still, forgetting war in Middle-
earth; for their wailing voices spoke to me of the Sea. The Sea! . . .
deep in the hearts of all my kindred lies the sea-longing, which it
is perilous to stir. Alas! for the gulls. No peace shall I have again
under beech or under elm.' Or as Frodo comments — when Sam
says of Rivendell that 'There's something of everything here' —
'Yes . . . except the Sea.' And the final end of the Fellowship of the
Ring in Middle-earth comes on the shores of the Sea.

Actually, Tolkien sometimes uses stars to make the same point.
When Sam, looking up from the depths of despair in Mordor, saw
a star overhead, 'The beauty of it smote his heart, as he looked up
out of the forsaken land, and hope returned to him. For like a

shaft, clear and cold, the thought pierced him that in the end the Shadow was only a small and passing thing: there was light and high beauty for ever beyond its reach.' But the light of the stars too originates from beyond the shores of Middle-earth.

What connects Nature to the spiritual, or requires the presence of the latter? In positive terms, as Alkis Kontos points out, when nature was still largely experienced as integral, alive and active, 'It was the *spiritual* dimension of the world, its enchanted, magical quality that rendered it infinite, not amenable to complete calculability; spirit could not be quantified; it permitted and invited mythologization.' And, I would add, it still is and does.

Negatively, Tolkien himself put it this way, in the course of discussing 'escapism': 'There are other things more grim and terrible to fly from than the noise, stench, ruthlessness, and extravagance of the internal-combustion engine. There are hunger, thirst, poverty, pain, sorrow, injustice. . . . And lastly there is the oldest and deepest desire, the Great Escape: the Escape from Death.' The last is probably the most important, but I would first like to consider 'injustice.'

## The 'Problem' of Evil

Tolkien has received considerable criticism on this point. Thus, Robert Giddings:

> The evil in the world as portrayed by Tolkien has nothing whatever to do with social or economic causes. It is evil, pure and simple. Consequently there is no need for change of socio-economic conditions, the environmental conditions of life, relations between different classes, etc., etc. — all these things which make up the very fabric of a society, of *any* society, are perceived by Tolkien as totally beyond any need or possibility of change.

Giddings exaggerates inexcusably — the *The Lord of the Rings* is full of social, economic and environmental changes which are directly related to the War of the Ring, and all its participants recognize their crucial effects. Many of them spend a great deal of their lives, and sometimes lose their lives, combatting evil as it exists in their world — hardly the actions of those who think they can change nothing, or only want a quiet life.

Nonetheless, Giddings is not altogether incorrect about Tolkien's position. As Gandalf repeatedly stresses, all one can do is combat evil when and where one is, and there is no permanent solution; indeed, the belief in final solutions is itself productive of the most grotesque evil. In the wizard's words, 'it is not our part to master all the tides of the world, but to do what is in us for the succour of those years wherein we are set, uprooting the evil in the fields that we know, so that those who live after may have clean earth to till. What weather they shall have is not ours to rule.' Ultimately, Tolkien is of the same opinion as Primo Levi: evil 'spreads like a contagion. It is foolish to think that human justice can eradicate it. It is an inexhaustible fount of evil . . .' Or, only slightly less darkly, William Empson, who described as 'one of the permanent truths' that 'it is only in degree that any improvement of society could prevent wastage of human powers; the waste even in a fortunate life, the isolation even in a life rich in intimacy, cannot but be felt deeply, and is the crucial feeling of tragedy.'

Ursula Le Guin, not for the first time, puts her finger on it: 'Those who fault Tolkien on the Problem of Evil are usually those who have an *answer* to the Problem of Evil — which he did not.' For such people, there is no real, intractable and ultimately non-negotiable evil, because, in Marina Warner's words, 'monsters are made, not given. And if monsters are made, they can be unmade, too.' Thus, interviewed about the recent slaughter of Scottish school-children, Warner delicately eschews 'evil' for 'vice'; even when confronted with the example of Nazism, she can only suggest that 'I don't think there was enough resistance there. People were duped or taken in and the *vitiation* spread . . .'

This is a serious failure of the moral imagination. It encourages us to believe that while grand social engineering — of which Stalin, Mao and Ceausescu's murderous attempts were only the most grandiose — may have failed, we should now put our faith in genetic engineering instead: straight from the frying-pan of 'nurture' into the fire of 'nature,' and anything, as usual, rather than self-knowledge.

Still worse than such dangerous superficiality, this particular liberal creed is highly irresponsible. It leaves those to whom it is taught cruelly exposed and unprepared for the realities of human suffering, especially that caused by other humans: as if, to quote Le Guin:

> evil were a problem, something that can be solved, that has an answer, like a problem in fifth grade arithmetic. . . . That is escapism, that posing evil as a 'problem,' instead of what it is: all the pain and suffering and waste and loss and injustice we will meet all our lives long, and must face and cope with over and over, and admit, and live with, in order to live human lives at all.

That is the view embodied in *The Lord of the Rings* too. Of course, the nature (or even existence) of evil is itself not something that comes with an answer in the back of the book, against which an opinion is 'right' or 'wrong.' Any response to evil is therefore inevitably problematic and incomplete; but Tolkien's kind is at least as complex and tenable as that of his more meliorist opponents. Besides, as D. J. Enright has noticed:

> No one entirely sure of himself, no one who disbelieved confidently in *evil*, who believed whole-heartedly in psychological or social explanations, would show quite such livid fury at the mere mention of the word as so many of us do.

## Death

There have been many trivializing attempts to pin down an overly-specific modern evil that can be identified with the Ring. I have noted and applauded Tolkien's resistance to any such move. In one letter he commented,

> Of course my story is not an allegory of Atomic power, but of *Power* (used for Domination). Nuclear physics can be used for that purpose. But they need not be. They need not be used at all. If there is any contemporary reference in my story at all it is to what seems to me the most widespread assumption of our time: that if a thing can be done, it must be done. This seems to me wholly false. . . . However, that is simple stuff, a contemporary & possibly passing and ephemeral problem. I do not think that even Power or Domination is the real centre of my story. . . . The real theme for me is about something much more permanent and difficult: Death and Immortality . . .

And again: 'There is I suppose applicability in my story to present times. But I should say, if asked, the tale is not really about Power and Dominion: that only sets the wheels going; it is about Death and the desire for deathlessness. Which is hardly more than to say it is a tale written by a Man!'

Part of his 'message' here, he once added, was 'the hideous peril of confusing true "immortality" with limitless serial longevity. Freedom from Time, and clinging to Time. . . . Compare the death of Aragorn with a Ringwraith.' The latter is the result of what Tolkien called 'endless serial living.' What a wonderful phrase! It encompasses both the increasingly fashionable practice of cryogenics — that is, freezing the head and/or body immediately after death, in the morbid hope of subsequent revival, thanks to the 'progress' of science — and the increasingly common practice of grotesque medical experiments to prolong and ease human life using involuntarily sacrificial animals, bred only to die for their organs or tissues to replace failing human equivalents.

Such is the modern understanding of immortality, typically literal-minded. There are no guarantees in Middle-earth of immortality, only glimpses and hints at best; but if immortality exists, it is clear that it does so only on the other side of death, not by trying to indefinitely extend life. The attempt to reject death — 'the gift of the One to Men' — is inseparable from the attempted denial of nature, the body, and ultimately life itself. Abandoning embodied life on Earth entirely to the 'rational rule' of the megamachine, it proves, as John Fowles has remarked, 'the gathering speed with which we are retreating into outer space from all other life on this planet.'

This is indeed just what the late Timothy Leary advocated, calling it S.M.I.L.E. (Space Migration, Intelligence Increase, Life Extension) — a suitably ghastly acronym that closely resembles C. S. Lewis's N.I.C.E. (National Institute for Co-ordinated Experiments), with a menacingly bright Californian spin. But the result is not immortality. Intervention in acute medical crises is one thing, but repeatedly deferring and delaying death by scientific and technological means produces, as Tolkien realized, only a population of Ringwraiths, taking forever to die:

> each 'Kind' has a natural span, integral to its biological and spiritual nature. This cannot really be *increased* qualitatively or quantitatively; so that prolongation in time is like stretching a wire out ever tauter, or 'spreading butter ever thinner' — it becomes an intolerable torment.

Another possibility is a memory-bank of ghosts. Here is Dr Chris Winter, head of the British Telecom Artificial Life project called 'Soul Catcher 2025,' enthusing about a microchip attached to your brain which could be used to recreate your experiences (*all* of them, unedited): 'This is the end of death, immortality in the truest sense. Future generations would be able to reconstruct you because your life, your soul, would live forever in silicon.' As he adds reassuringly, 'To use an analogy, we have split the atom but the bomb has not yet been built.'

Such pathetic simulacra of immortality have nothing in common with considered, willed and crafted artefacts like works of art and literature; their creators were never so confused that they thought of their work as a literal survival of death. Yet pathetic or not, various forms of electronic immortality *will* be taken seriously and touted about, because the money riding on them will see to it; and because our moral, cultural and educational impoverishment makes such 'success' possible.

Of course, it is one thing to assert and appreciate the profound value of limits, now even more unfashionable and crucial than when *The Lord of the Rings* was published, but quite another to do so when faced with the ultimate personal limit (so far as most of us know), personal death. Tolkien was very well aware of this, and in fact saw it as one of the keys to his beloved *Beowulf*. He called it:

> the theory of courage, which is the great contribution of early Northern literature. . . . The Northern gods . . . are on the right side, though it is not the side that wins. The winning side is Chaos and Unreason — mythologically, the monsters — but the gods, who are defeated, think that defeat no refutation. . . . It is the strength of the northern mythological imagination that it faced this problem, put the monsters in the centre, gave them victory but no honour, and found a potent but terrible solution in naked will and courage.

It is a mark of the importance Tolkien ascribed to this trait that it is fundamental to the hobbits, themselves fundamental to both *The Hobbit* and *The Lord of the Rings:* 'There is a seed of courage hidden (often deeply, it is true) in the heart of the fattest and most timid hobbit, waiting for some final and desperate danger to make it grow.' Without will and courage, Frodo would have surrendered the Ring to Sauron in Book One. Without 'luck' too, however, the end result would have been the same.

## Luck, Fate, Providence

Tolkien himself was 'a staunchly conservative Tridentine Roman Catholic . . . [with] a special reverence for the Virgin Mary.' This faith undoubtedly permitted considerably more room for femininity, intermediary deities, nature, and morally neutral or ambiguous spirits than a strict Protestantism would have countenanced. Even its very strength may have been what allowed him to remain on intimate terms with pre-Christian (and especially Old English) paganism. Certainly, as a Christian, Tolkien believed that the victory of the monsters was not final.

His books, however, are what chiefly concern us; and they are not so simple. For example, *The Lord of the Rings* contains repeated hints that (in Gandalf's words) there is 'more than one power at work' — another power, that is, beyond even that of the greatest in Middle-earth, namely Sauron. Thus Gandalf tells Frodo, 'I can put it no plainer than by saying that Bilbo was *meant* to find the Ring. . . . In which case you were also *meant* to have it.' And later he describes his fateful encounter with Thorin Oakenshield on 15 March 2941 T.A. ironically: 'A chance-meeting, as we say in Middle-earth.' The same ambiguous quality often clings to his use of the word 'luck.'

But as Shippey points out, 'Mordor and "the Shadow" are nearer and more visible than the Valar [or gods] or "luck." This lack of symmetry is moreover part of a basic denial of security throughout *The Lord of the Rings*. . . . [Otherwise] the characters' courage would look smaller; and courage is perhaps the strongest element in the Tolkienian synthesis of virtue.' Otherwise, too, most of us would find Middle-earth a much harder place to recognize from our own experience. All that naked courage can do is enable one to carry on, as it does Frodo and Sam. But 'While persistence offers no guarantees, it does give "luck" a chance to operate, through unknown allies or unknown weaknesses in the opposition.'

In the Northern pagan world which provided Tolkien with so much of his material and inspiration, luck was greatly valued. H.R.E. Davidson writes:

> There seems little doubt that one of the strong prevailing in-
> fluences . . . was a deeply-rooted belief in the desirability of
> luck, and the part played by luck in men's lives. In a danger-
> ous and uncertain world, luck was essential if one was to sur-
> vive. . . . It was realized that skill and outstanding gifts were
> not sufficient to protect a man if he was 'unlucky-looking,' as
> the Icelandic sagas expressed it.

This understanding was closely related to the Anglo-Saxon idea of 'wyrd' or fate, which was decidedly not a fixed and irrevo-cable future, but 'a steady, ongoing process, only fully completed at the end of a lifetime,' within which 'there is room for personal liberty and choice, limited indeed, but still available for use.'

Such an idea extends naturally into Christian providence and God's help. In both cases, however, there is no question of 'luck' or 'chance' interfering with the exercise of free will. Thus, at al-most any point in *The Lord of the Rings*, as is made quite clear, things could have gone disastrously wrong. (In Tolkien's *Letters*, there is a chilling exploration of what would have happened to Frodo if he had succeeded in fighting off Gollum and claiming the Ring on Mount Doom.) Indeed, the only thing that finally gave this divine 'power' the opportunity to intervene at the crucial last hurdle, when Frodo is standing 'on the brink of the chasm, at the very Crack of Doom . . . black against the glare, tense, erect,' was not only his and Sam's stubborn persistence; it was their free exer-cise, and Bilbo's before them, of 'Pity, and Mercy.' Without that, there would have been no Gollum; and as Frodo says, 'But for him, Sam, I could not have destroyed the Ring. The Quest would have been in vain, even at the bitter end.'

## A Christian Work?

'Pity and Mercy' thus hold an indispensable place in *The Lord of the Rings;* and they are undeniably Christian virtues. So is it therefore a Christian work? I think the answer is: yes, but not only that. There are obvious and central elements of Christianity present. Frodo takes on the almost hopeless task of destroying the Ring in a spirit of humility; he forgives and spares the life of Gollum, when Gollum comes into his power; he is pacifistic, and a non-combatant even in the Battle of Bywater; and his quest succeeds because in this crucial instance, it is spiritual strength, not physical stature or military prowess, that matters. There is also Gandalf's transformation from the Grey into the White, which, although Tolkien was not trying to portray Gandalf as Christ, nonetheless has unmistakable connotations of Christ's resurrection.

Notice, however, that the destruction of the Ring and of Sauron, although it may have been made possible by a higher spiritual power, and perhaps God, was not *caused* by it; that would have transgressed both the Christian and the pagan versions of 'free will.' The triumph of the good is therefore shot through with contingency. It obeys Tolkien's prescription for fairy-tales of 'the Consolation of the Happy Ending,' which involves a 'sudden joyous "turn" . . . a sudden and miraculous grace.' But it is equally important that by the same token, this grace is 'never to be counted on to recur.' Thus as Kath Filmer has remarked, 'this consolation is not that of *faith* as might be supposed; it is the consolation of *hope* . . .' I believe this is entirely correct, and that what readers are finding in Tolkien's pages is not faith, but precisely, and above all, hope: something to which I shall return.

Nor is Tolkien's 'Happy Ending' final, for as he also says, 'there is no true end to any fairy-tale.' Thus, victory in *The Lord of the Rings* is but a temporary respite. Despite ambiguous hints about other worlds, Tolkien's book ends *in this one:* at the Grey Havens, where, after the departure of Frodo and Gandalf, Sam 'stood far into the night, hearing only the sigh and murmur of the waves on the shores of Middle-earth . . .' Even where the Ringbearers have gone is not a final heavenly realm guaranteeing immortality, but a special resting place:

> Frodo was sent or allowed to pass over the Sea to heal him — if that could be done, *before he died.* He would eventually have to 'pass away': no mortal could, or can, abide for ever on earth, or within Time. So he went both to a purgatory and to a reward, for a while: a period of reflection and peace . . . spent still in Time amid the natural beauty of 'Arda unmarred,' the Earth unspoiled by evil.

As his own definition of Recovery implies, Tolkien's 'evangelium' permits only a 'fleeting glimpse of Joy' in this world, not permanent transportation to the next.

Tolkien himself described Middle-earth as a 'world of "natural theology",' containing 'a monotheistic but "sub-creational" mythology.' Writing to Fr Robert Murray, he maintained that *The Lord of the Rings* 'is of course a fundamentally religious and Catholic work; unconsciously so at first, but consciously in the revision. That is why I have not put in, or have cut out, practically all references to anything like "religion," to cults or practices, in the imaginary world. For the religious element is absorbed into the story and the symbolism.' Now it is a curious and important question why Tolkien should have *wanted* to cut out all references to religion in 'a fundamentally religious work'; we shall return to it. First, however, I want to contest Tolkien's own description of *The Lord of the Rings* here — despite the undeniable presence of

Christianity in it — as economical with the truth; or at least, inadequate.

## A Pagan Work?

What of the 'natural theology' of Middle-earth? True, it is nominally monotheistic. At the top is God, called 'the One.' Below Him is a pantheon of gods and goddesses called the Valar. As Tolkien admits, however, He 'indeed remains remote, outside the World, and only directly accessible to the Valar or Rulers;' these 'take the imaginative but not the theological place of "gods".' But of course, *The Lord of the Rings* is an 'imaginary,' not a theological, text. And in it, the One only directly intervened in history once, in the momentous reshaping of the world in the Second Age. There is never the slightest suggestion that He would do so again, no matter how badly matters went in the War of the Ring.

The Valar, also described as 'the Guardians of the World' and, significantly, 'Powers,' are more present. They have at least visited Middle-earth, and one in particular — Elbereth — is the object of song, prayer and supplication in *The Lord of the Rings*. Furthermore, they are related to the ancient elements (fire, earth, air and water) in a characteristically pagan way. All this, it seems to me, introduces a real element of pagan polytheism into the picture.

Other aspects of Tolkien's work point to the same conclusion. For example, there is much evidence of an active animism, a natural world that is literally alive. In *The Hobbit*, everything is bumped up a level, so to speak: the Lonely Mountain has roots, while the roots of trees are 'feet.' In *The Lord of the Rings*, the mountain Caradhras shows his displeasure by snowing heavily to block the Company's way; the herb *athelas* makes the air sparkle with joy; Sauron's attack is reflected in great engulfing clouds, and the subsequent change in the winds prefigures the turn of the tide in the battle for Minas Tirith. This, and much else, is contained in one of Tolkien's most marvellous passages, when the Captain of the Nazgûl confronts Gandalf before the ruined gates of Minas Tirith:

> in that very moment, away behind in some courtyard of the City, a cock crowed. Shrill and clear he crowed, recking nothing of wizardry or war, welcoming only the morning that in the sky far above the shadows of death was coming with the dawn.
>
> And as if in answer there came from far away another note. Horns, horns, horns. Great horns of the North wildly blowing. Rohan had come at last.

Again, after the battle, 'A great rain came out of the Sea, and it seemed that all things wept for Théoden and Éowyn, quenching the fires in the City with grey tears.' The 'as if' and 'it seemed' here are plainly a sop to modern rationalists, and when Tolkien writes, 'Tree and stone, blade and leaf were listening,' he does *not* mean it metaphorically.

Equally, the blasted and poisoned landscape around Mordor is as much evidence of Sauron's moral nullity as it is ecological commentary. For Tolkien, as for Ruskin, the signs of the sky and earth were literally the signs of the times: '"Blanched sun, — blighted grass, — blinded man",' together constituted 'a moral as well as meteorological phenomenon: it was a blasphemy against nature . . .'

Polytheism and animism are, of course, 'pagan' by definition; and the celebrations of 1420 T.A. were a veritable pagan feast (one could almost say 'orgy'). On Midsummer Eve — not just any old day in the year — 'the sky was blue as sapphire and white stars opened in the East, but the West was still golden, and the air was cool and fragrant . . .' This is the setting for the symbolic marriage (and its subsequent consummation) of the King and his bride, Arwen Evenstar. It comes as no surprise that 1420 became famous for its weddings, and in an inverse 'Wasteland' effect the land too was restored to fertility. As Tolkien puts it, in a passage also revealing his fine light touch:

> Not only was there wonderful sunshine and delicious rain, in due times and perfect measure, but there seemed something more: an air of richness and growth, and a gleam of a beauty beyond that of mortal summers that flicker and pass upon this Middle-earth. All the children born or begotten that year, and there were many, were fair to see and strong. . . . The fruit was so plentiful that young hobbits very nearly bathed in strawberries and cream; and later they sat on the lawns under the plum-trees and ate, until they had made piles of stones like small pyramids or the heaped skulls of a conqueror, and then they moved on. And no one was ill, and everyone was pleased, except those who had to mow the grass.

There are additional interesting complications in the religious and theological picture. Both Elves and Dwarves apparently believed in, or rather practised, reincarnation (although not necessarily of the same kind). In an impressive testimony to his open-mindedness, Tolkien defended this in reply to a Christian reader who felt he had 'over-stepped the mark in metaphysical matters,' saying: 'I do not see how even in the Primary World any theologian or philosopher, unless very much better informed about the relation of spirit and body than I believe anyone to be, could deny the *possibility* of re-incarnation as a mode of existence, prescribed for certain kinds of rational incarnate creatures.' And Tolkien was prepared to take other creative liberties with the received Christian wisdom; as Tom Shippey points out, Frodo leads himself into temptation but is delivered by evil. Divination, too, long a *bête noire* of the Church, figures too, in Galadriel's scrying pool. In addition, these things often have other and far older lineages than just their relatively recent Christian versions.

## Wizards and Stars

For example, let us consider Gandalf a little more closely: 'a bearded stranger seeming in long cloak larger than life,' 'an old wanderer glancing up from under a shadowy hood or floppy-brimmed hat . . . with a gleam of recognition out his one piercing eye,' whose chief skill was 'as a wizard or sorcerer or *vates*,' in his 'usual disguise of wide-brimmed hat, blue cloak, and tall staff.' He usually appeared as 'a tall, vigorous man, about fifty years of age . . . clad in a suit of grey, with a blue hood, and . . . a wide blue mantle flecked with grey . . . on his finger or arm he wore [a] marvellous ring . . .'

To anyone who knows the books, the description is unmistakable. Yet these are not Tolkien's words, and they were used to describe not Gandalf but Wodan (Odin in Norse), the chief god of the Old English pantheon. The same god sometimes competed with giants in tests of esoteric knowledge, encapsulated in riddles, and triumphed in the end by keeping his opponents 'so engrossed in the game of question and answer that they were caught by the rays of the rising sun and turned to stone' — exactly the trick Gandalf used on the trolls in *The Hobbit*.

His transformation into Gandalf the White notwithstanding, 'the Odinic wanderer,' as Tolkien once called him, is a profoundly pagan character, a mage and shaman, with parallels in every cultural memory: the Celtic Merlin and the classical Hermes Trismegistus, to name but two well-known ones. (The portrait of the latter on a paving-stone in Siena Cathedral could be of Gandalf himself.) And while Gandalf is neither *The Hobbit*'s nor *The Lord of the Ring*'s central character, equally they are unimaginable without him.

Then there is the matter of Eärendil. An Old English poem in the *Exeter Book* includes these words: 'Oh, Eärendel, brightest of angels, sent to men above Middle-earth . . .' (or alternatively, 'sent from God to men'). Tolkien brought this passage to the attention of Clyde Kilby, describing them as 'Cynewulf's words from which ultimately sprang the whole of my mythology.' Kilby then asks rhetorically of Tolkien's mythology, 'can we any longer doubt its profound Christian associations?' Well, we must certainly admit 'associations,' but they are far from exclusive ones. For whereas Eärendel was originally simply 'the old name of a star or planet,' Tolkien specifies it as the Morning and Evening Star, the brightest 'star' in the heavens — namely, Venus.

The associations surrounding Elbereth, Tolkien's fictional goddess of feminine compassion, point the same way. Her name translates as 'Star-lady' (alternatively, Elentari = 'Queen of the Stars,' or Varda = 'Lofty'). Through the millennia-old identification of the planet and the goddess, Elbereth's antecedents as pagan Aphrodite-Venus are again just as precise and powerful as those of the Christian Virgin Mary, and considerably older. Indeed, they are those of Mary herself, honoured as the Queen of the Heavens, and already ancient when Lucretius (*c.* 99–55 BC) praised Venus, in words that any Elf would have found perfectly acceptable — 'Thou alone, O goddess, rulest over the totality of nature; without thee nothing comes to the heavenly shores of light, nothing is joyful, nothing lovable.'

This permeation extends from such vital elements of Tolkien's literary myth through to the relatively trivial, if still enjoyable. Take the rather marginal wizard Radagast the Brown, for example, who has a special affinity with animals. It is hard to believe that he has nothing to do with Radegast, the pagan patron of the Beskyd Mountains in the Czech Republic, who appears in statues there in a horned helmet with a large bird sitting on him. (Radegast has also lent his name to a rather good Czech beer.)

It could even be said that Tolkien's religious mythology is, in one major respect, not 'supernatural' at all, but humanistic. Jack Zipes argues that one of Tolkien's intentions was 'to demonstrate how human beings could "recover" religion to offset the forces of inhumanity. The irony here is that Tolkien raises the small person, the Hobbit, to the position of God, that is, he stands at the centre of the universe. . . . God is absent from the [sic] Middle Earth. The spiritual world manifests itself through the actions of the redeemed small person.' And in this context, it is worth noting Tolkien's remark that 'The Hobbits are, of course, really meant to be a branch of the specifically *human* race (not Elves or Dwarves) . . .'

## All and None

None of this is intended to deny or denigrate the Christian elements in Tolkien's work. In particular, as Patrick Grant has pointed out, 'the concept of Christian heroism, a spiritual quality that depends on obedience rather than prowess or personal power,' is an integral part of *The Lord of the Rings* in the person of Frodo. 'The spiritual interpretation of heroism,' he adds, 'is the most significant Christian modification of the epic tradition . . .' Indeed, none of the strands I have identified should be taken as somehow trumping or cancelling out the others. I am not suggesting, for example, that *The Lord of the Rings* is either 'really' or 'unconsciously' pagan; Tolkien himself was rightly dubious about unreconstructed paganism, although he did have a lot of time for 'pagan virtues' such as courage.

More generally, in any case, I agree with Sean Kane that, for better or worse, 'The gods have not been silenced; in fact, they have been driven underground;' and Russell Hoban in the same vein:

> gods do not replace one another. Let prophets and kings do what they will: gods are a cumulative projection of everything in us. I'm not trying to reduce this to psychiatry — I mean that we worship the gods projected by the god-force that projects us as well on the screen of its mind.

Or as he also says, 'We make fiction because we *are* fiction. . . . We make stories because we are story.'

Now, Tolkien's almost identical statement — 'We make still by the law in which we're made' — assumes a theistic Creator who made us. But the logic of what he is saying, it seems to me, can be accommodated by Hoban's 'god-force' without any significant loss. Whether Gods or people, or countless other beings, we are all in this together. If that is true, then his essential point here does not depend on a Judaeo-Christian God for either its coherence or force. (I am very aware, of course, that Tolkien himself would probably have sharply disagreed with me on this point. But I am making a reasoned assertion about his book, and on that subject not even its author is the sole or final arbiter.)

Adherence to any major world religion involves a vital awareness of something that isn't a thing, that is always more than us. It allows us to recognize that all knowledge is suffused and delimited by mystery, and all initiative by dependence.

Against this great (potential) good, however, must be balanced several problems with Tolkien's mythic Christian theism, which identifies the life of Christ as the *unique* point where history and myth coincide. (Please note that I am not speaking theologically here, but rather of the way Christianity has been institutionalized and commonly interpreted; in short, its social and historical effects.)

One problem is not that the Christian message is too radical, but that it does not go far enough. A human life as the coincidence of mythic truth and historical reality is presented as restricted to one person who lived 2000 years ago, and therefore a vicarious one at best for the rest of us. Now, this may be a theological misunderstanding; but it is a very widespread and longstanding one. The only remedy may well be to declare such an experience, explicitly and fully, as open, in principle, to everyone of and for *themselves*. As P. L. Travers put it: 'in the long run, whatever it may be, every man must become the hero of his own story; his own fairy tale, if you like, a real fairy tale.'

Another and more serious problem follows on from this point. The experience of another person's life in such a way is something irreducibly personal, and cannot authentically be imposed on others. Again, I am concerned with the way Christian doctrine has been and is commonly taken. If Christ — or anyone else, whether incarnation, prophet or God — is held to be the first, last, and only exemplar of religious truth, then this particular version becomes not only true or useful or good (which may indeed be the case) but necessarily and universally so, to the ultimate exclusion of all other such stories, events and people. In that case, either most of the world (let alone the non-human world) must be abandoned to error, or else a campaign of conversion is demanded to *enforce* 'universal' truth. Exclusivity is bad enough, but the consequences of aggressive inclusivity have been unfailingly horrible.

Finally, as we shall see, those consequences are not limited to the human world: given belief in a (single) transcendental Creator, outside and above His creation, it is impossible for the Earth, its places and beings to have primary or ultimate value.

*The Lord of the Rings* transcends any strictly monotheistic reading. Instead, it manifests an extraordinary ethico-religious richness and complexity which derives from the *blending* of Christian, pagan, and humanist ingredients. It is all of these, and no single one of them. They can be separated analytically, of course, but not their joint and mutual effects — any more than can the different flavours that make up a soup. And when we turn from such internal considerations to how and why Tolkien wrote what he did, the point emerges clearly that the work's syncretism, including (indeed requiring) the elimination of 'practically all references to anything like religion' (as we now understand it) was a conscious and deliberate decision, and a very wise one.

## Post-Christian/Neo-Pagan/New Times

The clue to Tolkien's decision lies in his old exemplar, the author of *Beowulf*. In his major British Academy lecture, Tolkien characterized the poem as 'a fusion that has occurred *at a given point* of contact between old and new, a product of thought and deep emotion.' Living in such a time, when paganism (including its 'Northern courage') was succumbing to the new religion — although unevenly and unpredictably — its author had responded to this dilemma by suppressing 'the specifically Christian.' Is it surprising, then, that Tolkien should decide to emulate the Beowulf-poet, and see to it that 'the religious element is absorbed into the story and the symbolism'? For Tolkien was keenly aware that he too lived 'at a given point': the other end of the same historical epoch, the 'post-Christian' (as well as postmodern) to *Beowulf*'s 'pre,' when once again there is no single clear and over-arching set of values. All different religious and secular values mix and collide; there is no single criterion by which to judge between them that is even nearly universally accepted, yet none of them is unaffected by the others.

For that reason one cannot meaningfully speak of a collective 'return' to any single one of them. Indeed, what are contemporary fundamentalism, nationalism and racism but a distorted and hopeless response to an all-pervasive 'basic denial of security'? If there is one dictum that sums all this up and suggests a non-neurotic response, it must be Joseph Schumpeter's advice: 'To realize the relative validity of one's convictions and yet stand for them unflinchingly is what distinguishes a civilized man from a barbarian.' And it is entirely fitting that it is a civilized man, instead of the barbarians, who now embodies the pagan virtue of 'Northern courage' that Tolkien so admired. (As part of the same process, the green of abundant foliage which used to signify barbarism is now becoming the sign of a society that is sufficiently civilized to value nature.)

Thus, Tolkien's story combines Christian humility and mercy — pre-eminently, Frodo — with a pagan appreciation of places and powers — Gandalf, Lothlórien, and many others — and humanist virtue, in the hobbits: ordinary, small people whose contribution becomes crucial. Such a combination is not the only possible one; it takes no account of Judaism, nor of Eastern religions. But it is a shrewd and hopeful synthesis, which emphasizes the chief virtues of each of its three elements. It also recognizes that 'reason' alone will never suffice to save what is rare and fair, both human and natural, in this world. Arguments from pure utility have already conceded the central ground to the forces of destruction. For reasons already discussed in the previous chapter, the things, places and people we love will be saved for their own sakes or not at all; and that is ultimately a religious valuing.

Some may see this typically postmodern flux and mingling as decadence. It can also be interpreted hopefully, however, particularly if you think (as I do) that no single tradition holds the key to resolving our contemporary ills. For example, like every other major world religion, Christianity has 'green' possibilities which hold out some hope in a global ecological crisis. The Bible can be seen as encouraging responsible stewardship of nature, and while this falls well short of denying nature its own autonomy and validity, it is also clearly preferable to ruthless exploitation for profit.

This positive possibility is reflected in a few specific strands: the Benedictine monastic tradition, for instance, and the example of St Francis. Celtic Christianity, to take another, invokes a monastic spirituality, living simply and lightly on the Earth and close to nature, that undoubtedly extends much earlier pagan and animistic cultural emphases. In the words of a Celtic blessing,

> *Deep peace of the running wave to you*
> *Deep peace of the flowing air to you*
> *Deep peace of the quiet air to you*

*Deep peace of the shining stars to you*
*Deep peace of the Son of Peace to you*

It would take a hard heart to deny the healing beauty of such an attitude, and how much it offers to individual seekers after understanding.

The trouble is, however, the extent to which such sympathies are outweighed and overpowered by the principal thrust of institutionalized Christian tradition (just as Celtic Christianity was, historically, by Roman Catholicism). Notwithstanding the theological evidence for more complex and subtle judgements, what continues to resonate most strongly, collectively, socially and historically, is the licence to exploit nature for human purposes stated in the very founding chapter of the Bible (Genesis 1:26–28), and augmented in Acts 10:11–13, when God tells Peter to 'Kill, and eat.' Essentially, God is seen removed from the world and nature, which paves the way very nicely to its desacralization; then, He briefly returns, but only in human form, to 'save' other humans. As Sean Kane truly writes, 'all the work that various peoples have done — all the work that peoples must do — to live with the Earth on the Earth's terms is pre-empted by the dream of transcendence.'

Furthermore, as I have mentioned, the monotheism of Judaeo-Christianity and Islam demands a single and universal God, which in turn requires a priesthood of licenced interpreters of His will, and thus creates the categories of heresy, apostasy, and so on. This logic went straight into modern science, which substituted scientifically factual truth for God; and economics, with its universal 'bottom line' of financial viability and profit. What suffers in all cases is the same: pluralism, local distinctiveness, and the unique. So too (as Max Weber realized) does enchantment, as a single god establishes a monopoly of meaning which implies that everything can be subjected to a single, 'rational' ordering.

This is an immense part of the contemporary social, ecological and spiritual crisis which 'returning to God' does nothing whatsoever to help; quite the contrary, wherever such faith slides into the fundamentalism to which it is inherently prone. Conversely, the spiritual dimension of life is decidedly not identical to formal and established religion.

To some, understandably, paganism seems to offer more obvious resources to meet our contemporary crisis. It is not monotheistic and exclusive: pagans tend overwhelmingly to be polytheists, animists, pantheists, even agnostics or atheists. By the same token, it is not heavily institutionalized: you can be any or none of these without fear of excommunication (at best). Pagans thus make up not a religion so much as a 'collective spirituality.' And at its heart is a love of and concern for the Earth and all its life-forms (not just the human): 'there is an "elective affinity" between Paganism and ecology. . . . Paganism is an ecological spirituality.' This also corresponds to the ecological wisdom of indigenous people, to many of whom it appears that Christianity 'only looks at people, whereas we look at all things and see them as our relatives. We see a spirit in all things.'

Yet a 'return' to these older ways is also not unproblematic. Paganism is itself a Christian term and partly its creation, and any reconstruction of it involves a great deal of guesswork and improvization. That may be just as well, for along with its 'positive' and 'nurturing' aspects, pagan religions of the past sometimes involved extraordinary cruelty and a deeply-mired darkness that is (or ought to be) unacceptable to us now. Nor is humanism, in itself, the answer. It still preserves a valuable spark of human initiative and qualified independence, but has now become bloated into an arrogant philosophy licencing any barbarism for any human ends, up to and including the destruction of any other life-forms (except those found to be 'useful') and indeed the living Earth itself.

## From Religion to Myth to Fantasy

Tolkien was horrified by the modern desecration of mankind and nature, along with the religious values that he saw as their last protection from that 'idolatry of artefacts' which his colleague and friend C. S. Lewis called the 'great corporate sin of our own civilisation.' Nonetheless, Tolkien 'realized the absurdity in post-Christian days' (in Richard L. Purtill's words) 'of attempting original myth.' His solution was a creation embodied in literary myth. (It could be said that *The Silmarillion* was precisely his attempt at 'original myth,' both for and within Middle-earth; and the result was not entirely a happy one.)

Zipes makes essentially the same point. 'Tolkien was acutely aware,' he writes, 'whether he stated this or not, that the essence of Christianity could only be conveyed to human beings in secularized form, given the changing referential framework of values in a capitalist world which has smashed the aura of the Judaeo-Christian tradition. . . . Thus, fantasy is not only art for Tolkien, but *religion*, secularized religion, which is informed by a chiliastic perspective of a redeemed humanity.' I think Randel Helms is still closer to the mark when he says that 'The poetry of the mythic imagination will not, for Tolkien, replace religion so much as *make it possible*, putting imaginatively starved modern man once again into awed and reverent contact with a living universe.'

Actually, Tolkien did state almost exactly what Zipes only infers, in some remarks in a letter about the Arthurian myth-stories. His chief objection to it was 'it is involved in, and explicitly contains the Christian religion. For reasons which I will not elaborate, that seems to me fatal. Myth and fairy-story must, as all art, reflect and contain in solution elements of moral and religious truth (or error), but not explicit, not in the known form of the primary "real" world. (I am speaking, of course, of our present situation, not of ancient pagan, pre-Christian days . . .)' Thus Tolkien needed Frodo and the hobbits not only to give his disabused mod-

ern readers access to the ancient heroic world of Middle-earth —
'Dear me! We Tooks and Brandybucks, we can't live long on the
heights' — but also as a mediation, like *The Lord of the Rings* as a
whole, 'between pagan myth and Christian truth.'

With its usual extraordinary detail and consistency, this point
is even implied within the world of *The Lord of the Rings*. For already in Frodo's day, 'Gone was the "mythological" time when
Valinor (or Valimar), the Land of the Valar (gods if you will)
existed physically in the Uttermost West, or the Eldaic (Elvish)
immortal Isle of Eressëa; or the Great Isle of Westernesse
(Númenor-Atlantis).' These places, and what they represented,
were no longer available:

> Men may sail now West, if they will, as far as they may, and
> come no nearer to Valinor or the Blessed Realm, but return
> only into the east and so back again; for the world is round,
> and finite, and a circle inescapable — save by death. Only
> the 'immortals,' the lingering Elves, may still if they will,
> wearying of the circle of the world, take ship and find the
> 'straight way,' and come to the ancient or True West, and be
> at peace.

Was this supposed to have been a 'real' shift from a geocentric
disc-world to a heliocentric globe, or simply a shift in cosmologi-
cal perception? The question is misplaced; in the books' true pre-
(and post-) modern style, reality and perception cannot meaningfully be separated. In either case, the effect is the same: the seas are
now bent and the old 'straight way' gone, and with it all 'straight
sight.' So it is in our day too. We have only a few leaves, such as
Tolkien's, from the great tree of the Earth's wisdom. But that does
not devalue them; quite the reverse.

# FANTASY, LITERATURE AND
# THE MYTHOPOEIC IMAGINATION

---

*The lover of myth is in a sense a lover of
Wisdom, for myth is composed of wonders.*

---

SO FAR WE HAVE considered community, nature and the spirit
in Tolkien's principal work. But the Shire, Middle-earth and the
Sea can only be reached through the pages of a book, a literary
artefact, and a very particular kind at that. It is time to see what
kind.

You will remember that Tolkien had once wanted to produce
an English 'mythology.' But this desire coexisted and sometimes
conflicted with others, reflected in concerns that I have already
discussed, plus the demands of simply telling and sustaining such
a long story, 'a tale that grew in the telling' to include more and
more. Tolkien's initial ambition had serious problems, in any case;
as a baffled early reader's response to a draft of *The Silmarillion*
demonstrated (he decided it must be Celtic), Anglo-Saxon
Englishness in the national psyche proved to be so unconscious
and muddled that creating a mythology for England 'was trying to
give people something which was so evidently missing that they
would not believe it if you gave it to them.' In the end, it was
partly abandoned and partly subsumed by Tolkien's other and

larger concerns: 'The pull of faërie was stronger than the pull of Old English.' And since the counterpoint to his literary leaf of Faërie was modernity itself, this creative expansion was complemented by the way his books have become effectively universal in their reception, with many non-English readers far from 'the North-West of the Old World' hearing Tolkien speak directly to them and their lives.

The form that Tolkien hit upon as best embodying what he wanted to say was an amalgam of myth and fairy- or folk-tale, themselves close relations, reworked to accommodate contemporary readers via his own creation, the hobbits. But that raised awkward questions too: if the mythological days are over, the days when gods and spirits walked and talked with humans — or were recognized as doing so, which amounts to the same thing — then how can authentic myth be created?

Tolkien realized, as I have already suggested, that it can't. It has to be *re*-created, in the form of a contemporary literary myth or mythopoeic fiction. While the result still partakes of myth — how could it not, when it was in his metaphysical bones? — it also includes sufficient elements of the realistic and secular novel to provide access to modern readers, and thus to enable 'these old things' to survive in a hostile literary milieu.

I shall once again use my own experiences as a reader as an entry-point into these complex issues. But first, let me clear the ground of one common and unnecessary misunderstanding: fairy-stories are not at all necessarily stories about fairies, but 'stories about Fairy, that is *Faërie*, the realm or state in which fairies have their being.' Faërie thus also contains *us*, when we are enchanted. Let me also remind you that according to Tolkien, fairy-stories offer, 'in a peculiar degree or mode, these things: Fantasy, Recovery, Escape, Consolation . . .'

I first read *The Lord of the Rings* at the age of sixteen, after an unwitting preparation thanks to *The Hobbit*, seven years earlier. (And yes, perhaps it has stayed with me partly from this early exposure, before a more developed critical sense could bar the way: but does that make Orwell and Huxley, whom I also first read at the same age, adolescent writers?) I was overcome from the beginning by the unmistakable sense of having encountered a world that was more real than the one I was then living in, or the reality of which was much more concentrated. Accompanying this feeling was the equally odd one of inexplicable familiarity with that world. And finally, there was a definite sense of loss when I had finished, which, combined with delight and curiosity, impelled me immediately to recommence reading it. None of this was a unique experience on my part; to a greater or lesser degree, *The Lord of the Rings* has affected many readers in just these ways, and they deserve some attempt at understanding.

## *Loss and Consolation*

Let's look at the sense of loss first. It is actually well-described within the book itself, at the point where the Company is setting off in their boats on the Silverlode: 'For so it seemed to them: Lórien was slipping backwards, like a bright ship masted with enchanted trees, sailing on to forgotten shores, while they sat helpless upon the margin of the grey and leafless world.' This feeling of deprivation, and the unwilling return to a 'grey and leafless world' upon finishing the book, is distinguishable from mere sentimentality by its inexplicable and sometimes painfully sharp poignancy. In the same vein — and it is no coincidence that this incident also involves Lothlórien, 'the heart of Elvendom on earth,' and by that token its ultimate repository of enchantment — when Frodo was walking up Cerin Amroth, he felt that 'When he had gone and passed again into the outer world, still Frodo the wanderer from the Shire would walk there . . .' Karen Blixen's masterpiece, *Out of Africa*, evokes the same kind of feeling; the only difference is that the enchanted land slipping away is the Ngong

Hills of Kenya, to be replaced by a disenchanted but inexorable Europe. (Interestingly, Blixen also strongly shared with Tolkien a respect for fairy-tales and an unmistakably 'Northern air.')

That passage concerning Frodo might remind one of another, from the glory days of Edwardian children's literature: 'wherever they go, and whatever happens to them on the way, in that enchanted place on the top of the Forest a little boy and his bear will always be playing.' And if, Dear Reader, that sort of thing makes you (like Dorothy Parker) want to fwow up, Tolkien's is less likely to. It is embedded in a much more sombre and distinctly adult view of life, with the monsters very nearly if not quite dead-centre; and as we saw in the previous chapter, the nostalgia it engenders is finally redirected back into our own lives here and now.

It is true that many readers of *The Lord of the Rings* are younger than the average readers of, say, literary novels. But it does not follow that the book is only for the young. It seems to me that they are likely to be more receptive to the wisdom that Tolkien offers, reading when so many questions about life loom so large, and they haven't already made up their minds about the answers. As Derek Brewer has shown, Tolkien skilfully exploits our ability as readers to turn space — archetypally a journey or quest — into time, thus following the development of the young 'Halflings' as they leave 'the cosy domesticity of the Shire' for a much bigger world where nowhere is really safe,

> where strange men much taller and older than oneself, facetious and impenetrable, offer help; and where one commits errors by sheer carelessness and inattention, . . . where . . . the men [are] admirable or evil; and where there are really rough characters . . .

These 'generalized psychological images of the personal human situation' extend from the passage of adolescence to the confrontation with death; and 'apart from the unduly fortunate careers of the Companions, the romance seriously imagines the brave and necessary confrontation of death with great variety and power.' (True, there is almost nothing about sex; but then, as Kim Taplin remarked, 'Everyone told us about sex . . .')

Edward Blishen has also well described a persistent feeling about the town and the country, in which the former was all maturity and responsibility, whereas 'the country was all "delight and beauty",' which thus coincided in some fashion with childhood itself, and that thereafter 'it must either be destroyed or, paradoxically, be for ever on the brink of destruction.'

This kind of awareness unites an 'inner' personal and spiritual dimension with an 'external' social and historical one; neither is predominant, and they are not naturally separable. Clyde Kilby reports that when asked why *The Lord of the Rings* should be so popular,

> I said I thought our present world had been so drained of elemental qualities such as the numinous, the supernatural and the wonderful that it had been consequently drained of much — perhaps most — of its natural and religious meaning. Someone wrote me of a sixth-grade pupil who, after reading *The Lord of the Rings*, had cried for two days. I think it must have been a cry for life and meaning and joy from the wasteland which had somehow already managed to capture this boy.

Thus the personal and psychological dimension of nostalgia extends imperceptibly into a more social and general awareness of a human 'loss of innocence,' and in consequence the kind of longing for Home we discussed earlier, and the social and even political value of which, following Fraser Harrison, I have tried to rescue. That longing for a world at peace with itself can have powerfully positive effects, as well as the negative ones that intellectuals usually perceive.

In my view, Tolkien's work awakes precisely such a longing. When it is keenly felt, 'pain and delight flow together and tears are the very wine of blessedness.' And for this, being a boy or girl is neither a problem nor a prerequisite. For those of any age with open minds and, more importantly, open hearts, Tolkien's practical and moral wisdom remains permanently on offer. Like the Grimm Brothers, he 'worked not for some prettified ideal of childhood but for the ongoing "What if?" that is humankind.'

Italo Calvino, in a profound passage, summed up the value of the kind of traditional folk-tales of which *The Hobbit* and *The Lord of the Rings* are an extended contemporary development:

> these folk stories are the catalog of the potential destinies of men and women, especially for that stage in life when destiny is formed, i.e., youth, beginning with birth, which itself often foreshadows the future; then the departure from home, and, finally, through the trials of growing up, the attainment of maturity and the proof of one's humanity. This sketch, although summary, encompasses everything: the arbitrary division of humans, albeit in essence equal, into kings and poor people; the persecution of the innocent and their subsequent vindication, which are the terms inherent in every life; love unrecognized when first encountered and then no sooner experienced than lost; the common fate of subjection to spells, or having one's existence predetermined by complex and unknown forces. This complexity pervades one's entire existence and forces one to struggle to free oneself, to determine one's own fate; at the same time we can liberate ourselves only if we liberate other people, for this is a *sine qua non* of one's own liberation. There must be fidelity to a goal and purity of heart, values fundamental to salvation and triumph. There must also be beauty, a sign of grace that can be masked by the humble, ugly guise of a frog; and above all, there must be present the infinite possibilities of mutation, the unifying element in everything: men, beasts, plants, things.

## Myth

What about readers' sense of hyper-reality? How could one feel intimately familiar, upon the very first reading, with an apparently imaginary place and/or time? Here I think mythology, although often loosely and over-used in connection with Tolkien, can be very helpful. A close relation of the folk- and fairy-tale, myth once signified narrative stories concerning the gods, as exemplified in the Western tradition by Homer. As early as Plato, however — but only achieving relatively complete dominance with the eighteenth-century Enlightenment — myth came to mean merely fabulous or unlikely fiction, near cousin to 'poetic truth,' or 'lies breathed through silver.' But such a definition concedes far too much to a scientistic rationalism/realism that as a philosophy is dubious, and as a practice is handmaiden to the destruction of community, nature and ethics alike. We can, and must, do better.

Carl Kerenyi defined the stuff of mythology as:

> an immemorial and traditional body of material contained in tales about gods and god-like beings, heroic battles and journeys to the Underworld — *mythologem* is the best Greek word for them — tales already well-known but not un-amenable to further reshaping. Myth is the *movement* of this material. . . . In a true mythologem this meaning is not something that could be expressed just as well and just as fully in a non-mythological way.

Furthermore, he writes, 'Myth gives a ground, lays a foundation. It does not answer the question "why?" but "whence?"' and thus, where to from here? Journeys in space and time, and existential choices — all this fits Tolkien like a fine glove, and indeed, his popular success supplies a clear answer to Kerenyi's slightly plaintive question, 'is an immediate experience and enjoyment of mythology still in any sense possible?'

Similarly, Milton Scarborough has recently argued that:

> myth is neither true nor false *in a theoretical sense* but viable
> or not viable for the tasks (both theoretical and otherwise)
> which confront us. This viability is not determined in intel-
> lectual terms but in the very process of living, by whether or
> not one is energized, whether or not problems are being
> solved, whether or not life is integrated at a variety of levels,
> whether or not it is endowed with a significance that pulls
> one toward the future in hope. Viability is not determined in
> advance of inhabiting a myth . . .

It is noticeable that Tolkien's severest critics have all done just
that: judged his work on a theoretical basis, from well outside it.
Unlike his many readers, they don't seem to have once set foot in
Middle-earth.

## *Local Mythology*

Virginia Luling rightly points out that 'All mythologies are necessarily both universal and local: universal in their scope, because
they deal with the nature of things; local in point of view and
"temper," because they arise out of particular cultures. This ten-
sion is present in the mythology devised by Tolkien, since it is
about both the human condition in general, and deliberately made
specific to a certain part of the world.' Let us consider the 'local'
dimension first.

Tolkien's incorporation into *The Lord of the Rings* of the cultural
traditions of England and 'the North-West of the Old World,' especially their mythologies, was not just a calculated strategy (as I
hope I haven't been taken to imply); it was unavoidable for a man
so thoroughly steeped in them. But that authenticity gives his
work a tremendous advantage over that of others, possibly otherwise similar, especially since those traditions still live. Thus, if I

had been able to articulate my early experience of inexplicable familiarity adequately at the time, I would have said it felt not so much like a discovery as a *re*-discovery, a reconnection with a living tradition that I hadn't been aware of. It does not just embrace the myths attached to England as Blake's preternaturally 'green and pleasant land,' although that counts too. Jared Lobdell is perceptively right to say that insofar as languages are among 'the chief distinguishing marks of peoples . . . we who read it in English are, as English-speakers, the inheritors of Tolkien's English mythology. . . .' But it goes beyond that to take in the pre-English Indo-European history of all the people from this turbulent corner of the world. The result, it seems to me, is the same sense of wonder that Keats experienced and recorded in his poem 'On First Looking into Chapman's Homer,' which left him 'Silent, upon a peak in Darien.' That new vista was one of our classical cultural heritage. Tolkien has performed the same service (with infinitely less thanks), in relation to that of 'North-West Europe,' for later, larger and less 'literary' generations.

The reality of Middle-earth, then, is not so strange. Mythologies, like other kinds of cultural traditions, contain, in coded form, a great deal of feeling and emotion in the form of accumulated human experience: the collective memories, hopes, wishes, fears and dreams of entire communities across time. As Alan Garner has noted, 'the elements of myth work deeply and are powerful tools. Myth is not entertainment, but rather the crystallization of experience, and far from being escapist literature, fantasy is an intensification of reality.' Tolkien's books present a highly distilled and concentrated, if unavoidably selective, instance of just that. It may also be that in a world saturated with Graeco-Judaic traditions, it has come as a relief to many to be offered access to badly neglected Northern ones — particularly if they still somehow feel like 'ours,' or even 'us.'

This analysis accords with Tolkien's own experience of writing

*The Lord of the Rings*, in which (he reports) it 'grew,' he 'was drawn irresistibly' toward certain things, and 'discovery' felt much more the case than 'invention.' If one reacts as reader in the manner I have described, it is impossible to feel that this was mere rhetoric designed to enhance his creation. And Middle-earth was emphatically not just 'created,' at least in the fashionable modern sense of creation *ex nihilo*, as a mark of the much-vaunted virtue of 'originality.' It was a co-creation, in partnership with some very old and durable cultural materials. Of course, it would be unduly extravagant to say that he literally discovered Middle-earth. But neither did he simply invent it.

## Universal Myth

The truth of the matter, the middle way, lies in what Tolkien, in his essay on *Beowulf*, called 'the mythical mode of imagination.' That he was indeed working with 'the movement' of myth cannot be doubted. The following description by Roberto Calasso was certainly not written with Tolkien in mind, but it could not be bettered as such:

> The mythographer lives in a permanent state of chronological vertigo, which he pretends to want to resolve. But while on one table he puts generations and dynasties in order, like some old butler who knows the family history better than his masters, you can be sure that on another table the muddle is getting worse and the threads ever more entangled. No mythographer has ever managed to put his material together in a consistent sequence, yet all set out to impose order. In this, they have been faithful to the myth.
>
> The mythical gesture is a wave which, as it breaks, assumes a shape, the way dice form a number when we toss them. But, as the wave withdraws, the unvanquished complications swell in the undertow, and likewise the muddle and

the disorder from which the next mythical gesture will be
formed. So myth allows of no system. Indeed, when it first
came into being, system itself was no more than a flap on a
god's cloak, a minor bequest of Apollo.

Here was Tolkien at work on *The Silmarillion*, perpetually trying to resolve the apparently countless variations and inconsistencies, and failing, and delaying publication yet again. It finally appeared posthumously, of course, edited by his son Christopher.

Why can Tolkien's 'mythology' be described as universal? I
have already shown that it embodies an attack on unchecked modernity in all its worst aspects, and presents a world of community, nature and spiritual values that successfully, albeit barely,
struggles to survive such destruction. That world seems to be a
different one, with strange people and places; yet at the same
time, it is also recognizably ours. And because the processes of
rampant modernization — economic, political, cultural — are now
truly global, the potential appeal and relevance of Tolkien's attack and alternative are also effectively universal. This is a social
and historical development; there is nothing necessarily mystical
about it.

But his universality comes about in another way, too. For
the very terms of his critique are mythic; after all, that is ultimately the most (and perhaps even only) effective way to counter a
worldview which is rigidly rationalistic and scientistic. And there
is literally nowhere in the world without some native tradition of
a mythical way of relating to the world in which it is alive and
saturated with spiritual meaning — enchanted, in a word. Those
traditions may be deeply buried, but — like the gods they embody — they can still be revived by recognition. Tolkien's living
mythicity thus touches older memories still, which are effectively
shared by all humanity. As such, it is a powerful stimulus to re-enchantment.

Paradoxically, its power is all the more universal for being a precise portrait of a time and place that (in a literal sense) never was. Tolkien's tale thus partakes of the fairy-tale's quality of 'Once upon a time' — never but always, nowhere but everywhere. As Sallust, the ancient Roman historian, wrote in *Of Gods and of the World:* 'These things never happened, but are always.'

## Back to Myth

This understanding of myth requires a still older and broader understanding than, say, Kerenyi's. I think Sean Kane, going back to myth's Paleolithic origins, supplies one: 'Myths are not stories about the gods in the abstract; they are about "something mysterious," intelligent, invisible and whole.' And that something always comes back to nature. Thus, 'The proper subject of myth is the ideas and emotions of the earth.' That *includes* people, of course, as one kind of living thing among many. But it is certainly not restricted to humanity.

Actually, there is a perceptible split within myth between this older kind and the succeeding stories of the agriculturally-based Neolithic (which gave us the Celtic, Germanic and Greek legends), whereby the ancient earth-mysteries and vegetable goddesses tend to become overlaid with powerful, and distinctly more anthropocentric, sky-gods; but not wholly, and not without extracting concessions. It seems to me that Tolkien's *mythos* is balanced between, and includes, both kinds. The Valar, for example, are plainly gods, or 'powers' in human form; whereas the Ents are *not* people in tree-form, but trees that speak and walk — that is their very point, and wonder — while the Drúedain, whom Théoden and Merry encounter on the way to Minas Tirith, are the indigenous Paleolithic and Mesolithic hunter-gatherers themselves. Tolkien's Elves are balanced on the very fulcrum of this shift, humanoid but chthonic: of the Earth, and even if killed, returning to it eventually through reincarnation.

Tolkien personally found this positioning deeply uncomfortable. In the original cosmogonic myth of *The Silmarillion*, the Two Trees of Valinor, white Telperion and Laurelin the Golden, bore the Moon and the Sun just before they succumbed to Morgoth's poison. Beginning in the late 1950s, however, he was assailed by self-doubts as to 'the astronomically absurd business of the making of the Sun and Moon,' and tried to purify his *legendarium* by replacing it with an account more focussed on the gods, human in form if not in powers. But the vegetable birth of the luminaries was integrally related to other elements which he needed to retain, and Tolkien died without ever having succeeded in resolving the resulting inconsistencies. His son Christopher commented:

> I think it possible that it was the actual nature of this myth that led him finally to abandon it. It is in conception beautiful, and not absurd; but it is exceedingly 'primitive.' . . . [Tolkien's] grave and tranquil words cannot entirely suppress a sense that there emerges here an outcropping, as it were, uneroded, from an older level, more fantastic, more bizarre.

The Elves' humanoid/chthonic ambiguity points to something else that has long puzzled me (although, as so often, it concerns myself as a reader as much as it does *The Lord of the Rings* 'itself'). That is the strange feeling evoked by the appearance in the story of Glorfindel, one of the first Elves we encounter. To Frodo, sensitized by his knife-wound, 'it appeared that a white light was shining through the form and raiment of the rider, as if through a thin veil.' Moments later, wearing the Ring, his human and hobbit companions appear as mere shadowy forms, but Glorfindel as 'a shining figure of white light.' Later, Gandalf tells Frodo that the High Elves 'live at once in both worlds, and against both the Seen and the Unseen they have great power.'

Without contradicting his explanation, I would say that the

reason this appearance of a humanoid figure who shines can cause such wonder is this: the Elves, as a race of beings who exist both within Tolkien's world and outside it, in the worlds of cultural, historical and religious experience, survive from the 'mythological days' when humans (or beings in human form that we recognize as, in some nontrivial way, ourselves) were not just the passive Cartesian objects of light and sight, but were equally its subjects and agents. As such, the Elves remind us of ourselves before 'the Nothing' (as Michael Ende calls it) came to deaden, and our eyes, rather than being merely the receptors of light but also, in some more than narrowly metaphorical sense, emitted it. After all, we still rightly speak of someone's eyes, in moments of being fully alive, as shining. It happens particularly often with children, if they haven't already been captured by the monologic of a System (commercial, religious or whatever) and turned into little pseudo-adults.

## Other Approaches to Myth

Philip Wheelwright argued that the modern loss of myth-consciousness is 'the most devastating loss that humanity can suffer: for . . . myth-consciousness is the bond that unites men both with one another and with the unplumbed Mystery from which mankind is sprung and without reference to which the radical significance of things goes to pot. Now a world bereft of radical significance is not long tolerated; it leaves men radically unstable, so that they will seize at any myth or pseudomyth that is offered.' (G. K. Chesterton once similarly remarked that when people stop believing in God, they don't believe in nothing, but in anything.) If that is so, Tolkien is owed a huge debt of gratitude for his work in maintaining myth-consciousness, as he has. 'How horrid!' (atavistic, childish, irrational) I can hear the modernist Adults cry. But have they made such a good job of things? Perhaps it is finally time to conclude with Russell Hoban that 'we've tried the other way; we've tried making both things and people *It*, and we've seen the results.'

There is another way of looking at this question, however, which I would like to caution against, and that is the tendency to turn everything into psychology, whether Jungian, New Age or otherwise. Psychologizing feeds off, and tends to produce, the destructive modern myth of individualism. The sovereignty of individual judgement and private conscience is indeed a hard-won and precious thing. But the idea that it is possible or desirable to be a fully autonomous and separate being simply leads to fully privatized, isolated and rootless individuals, at the mercy of purveyors of universal and abstract forces — whether those of 'science,' 'the market,' or 'God.' Such individualism destroys the very things without which no creative mythopoeia is possible: respect for nature, including the body, love of locale and its uniqueness, and communities across time which create and recreate wisdom in creative interchange.

In any case, myth really needs no help from this source, which is so often a process of secularizing and sanitizing for mass public consumption. Roberto Calasso writes of the ancient Greek pantheon, 'No psychology since has ever gone beyond this; all we have done is invent, for those powers that act upon us, longer, more numerous, more awkward names, which are less effective, less closely aligned with the grain of our experience, whether that be pleasure or terror.' Similarly, as P. L. Travers says, 'the unconscious' loses nothing and gains much by being called, simply, 'the Unknown.' It is true that sensitive and useful psychological exploration of myth is possible. But too often, turning everything numinous into an 'archetype' is simply another way of maintaining the old humanist position of Top Dog, so to speak, whereby, say, 'a god like the Raven is not really a god — he is simply another expression of the universal human Id.' Sean Kane adds that, 'It seems remarkably species-chauvinistic of psychologists to reserve for the human psyche qualities that are found everywhere outside it.'

## Story

People often read *The Lord of the Rings* repeatedly, sometimes annually. I wonder if they aren't unconsciously enacting the myths and folk-tales, whose repetition, varying within certain limits, is so integral to their cultural role in oral traditions. Be that as it may, Tolkien's 'mythology' comes in a particular literary form. Whether defined as myth, epic, quest, saga, fairy-story, comedy or romance — and it has been, all of these — it is story-telling of a kind long unfashionable as an 'adult' genre; but more unfashionable among critics than readers, as some of the former are at last starting to realize. The fact is, to quote Nuala O'Faolain, that 'The language of highbrow criticism can only cope with a certain kind of fiction. It has no vocabulary with which to discuss a world where neither the individual nor the society is self-conscious, and the author pretends not to be either. . . . The ordinary reader is far ahead of the critics in ease with such a world.'

The magazine *Private Eye* once sneered that *The Lord of the Rings* appeals only to those 'with the mental age of a child — computer programmers, hippies and most Americans.' Many academic critics make basically the same charge; this version just doesn't bother with the heavy theoretical machinery. Amanda Craig has responded that 'The fact that Tolkien's world appeals to computer programmers is possibly less a sign that it is infantile than that he developed a hypnotic style and narrative which quickens the reluctantly literate as well as the devoutly bookish. Few writers in any century can claim the same.'

The pariah status of narrative is an ongoing problem. It was early and brilliantly analysed by Walter Benjamin. In an essay entitled 'The Storyteller,' whose every resonance applies to Tolkien, he noted that:

the art of storytelling is coming to an end. Less and less frequently do we encounter people with an ability to tell a tale properly. More and more often there is embarrassment all around when the wish to hear a story is expressed. . . . The art of storytelling is reaching its end because the epic side of truth, wisdom, is dying out . . . no event any longer comes to us without already being shot through with explanation. In other words, by now almost nothing that happens benefits storytelling; almost everything benefits information. Actually, it is half the art of storytelling to keep a story free from explanation as one reproduces it. . . . There is nothing that commends a story to memory more effectively than that chaste compactness which precludes psychological analysis. . . . A great storyteller will always be rooted in the people, primarily in a milieu of craftsmen. . . . The fairy tale, which to this day is the first tutor of children because it was the first tutor of mankind, secretly lives on in the story. The first true storyteller is, and will continue to be, the teller of fairy tales.

Benjamin described the novel as a huge step away from storytelling, with its roots in the oral tradition; the modern cult of information is yet another major removal. And let's face it, most of the novels are bad enough: permitted only 'that necessary degree of irony which is the sole form of "honesty" modern prose styles or conventions readily allow. . . . Unhappy with myth, wary of emotion, harried by empty political terminologies, scornful of "character," eager, it seems, to refine, redefine and narrow down the material until the works in question are about themselves, nothing else but themselves. Affirmation, no. Consolation, certainly not.' No wonder that, as Tolkien maintained, 'the "fairy-story" is really an adult genre, and one for which a starving audience exists.'

What Tolkien called the basis of narrative fantasy — 'a recognition of fact, but not a slavery to it' — chimes closely with a postmodern sensibility. Brian Attebery has suggested that 'Postmodernism is a return to story-telling in the belief that we can be sure of nothing but story.' He shrewdly adds that by comparison, modernist and realist criteria are ill-equipped to handle the success of 'Tolkien's perfectly sincere, perfectly impossible narrative.'

## *Fantasy*

But critics can be slow to catch up. Take the contemporary fairytale genre of fantasy, which Tolkien unintentionally created in large part, and which is almost completely marginalized in élite literary discourse. To the few examples that get reviewed at all, modernity's clerks still tend to react with the usual arrogant epithets: 'childish,' 'irrational' and so on. And even if I am right about Tolkien's essentially hopeful meaning, such people will only see, as D. J. Enright quotes a television listing, '"that awful cliché, a message of hope." Most of the rest of the evening's viewing has to do with murder, sexual problems, sick comedy. No mention of awful clichés there.' You may as well try to feed Gollum on *lembas*, the delicious and sustaining waybread of the Elves: '"Ach! No!" he spluttered. "You try to choke poor Sméagol. Dust and ashes, he can't eat that. He must starve".'

The fact is that modern criticism, journalism and publishing are unthinkingly and aggressively secular, cynical, snobbish and incestuous. As Max Weber put it, 'Specialists without spirit, sensualists without heart; this nullity imagines that it has attained a level of civilization never before achieved.' Karel Čapek, his contemporary in Prague, described the same situation:

> look how often the cultural world pronounces a sentence of annihilating rejection. How old-fashioned ideas, other people's views, or those of the habitués of a different literary café, are arrogantly dismissed out of hand. . . . This is vari-

ously called literary criticism, ideological struggle, a matter of principle, or the generation gap. In truth it is merely prickly intellectual exclusiveness running around looking for something to turn its nose up at. If your nose is in the air, though, you cannot see properly . . .

Nor can you read properly.

Is it really so difficult to appreciate both the novel and fantasy? Does the former really demand such complete and utter loyalty? Maybe it would help to see the novel as a literary form as a particular and contingent development of a much bigger and older tradition. And if that tradition is one of narrative, one of its stories' most persistent themes is the quest. John Fowles argues that 'Every novel since literary time began, since the epic of Gilgamesh, is a form of quest, or adventure.' W. H. Auden, discussing Tolkien's success, attributed 'the persistent appeal of the Quest as a literary form . . . to its validity as a symbolic description of our subjective personal experience of existence as historical.' Even Proust's *In Search of Lost Time* has been described as 'Perhaps the greatest quest novel of all,' one which 'takes the reader by degrees . . . to revelation of the truth.' (Perhaps that is one reason why, without suggesting that Tolkien was the same kind of writer, or as great a writer, I hold Proust's work in equal respect and affection as I do *The Lord of the Rings*.)

A quest is, of course, at the heart of both *The Hobbit* and *The Lord of the Rings*. In the first, it is a fairly traditional one: the hero, Bilbo, leaves his comfortable home in search of treasure. Along the way he has various adventures, ranging from the uncomfortable to life-threatening, and in the course of negotiating them, most notably a dragon, develops an adventurous part of himself hitherto unsuspected. There is a twist, however: Bilbo gives away his own major share of the treasure, the Arkenstone, to bring about peace between the Elves and Dwarves. A quest is central to *The Lord of the Rings*, too, but here, in a brilliant stroke, Tolkien has upended its usual form: Frodo's tortuous journey to Orodruin is not made to find and keep a treasure, but to return to its source, and thus destroy, a great menace.

What then is the prerequisite for letting down one's Grown-Up literary guard, enjoying good fantasy, and incidentally recognizing Tolkien's greatness? There is at least one, and the great Indologist Heinrich Zimmer points directly to it:

> The dilettante — Italian *dilettante* (present participle of the verb dilettare, 'to take delight in') — is one who takes delight in something. . . . The moment we abandon this dilettante attitude toward the images of folklore and myth and begin to feel certain about their proper interpretation (as professional comprehenders, handling the tool of an infallible method), we deprive ourselves of the quickening contact, the demonic and inspiring assault that is the effect of their intrinsic virtue. We forfeit our proper humility and open-mindedness before the unknown, and refuse to be instructed. . . . What they demand of us is not the monologue of the coroner's report, but the dialogue of a living conversation.

## The Lord of the Rings *as Fantasy*

How then do Tolkien's books in particular work? Drawing on the power of ancient Indo-European myth, in the form of a fantasy 'novel,' Tolkien's books invite the reader into a compelling and remarkably complete pre-modern world, saturated with corresponding earlier values, which therefore feels something like a lost home. They are just the values whose jeopardy we most now feel: relationships of respect with each other, and nature, and (for want of a better word) the spirit, which have not been stripped of personal integrity and responsibility and decanted into a soulless calculus of financial profit-and-loss. Wisdom in Middle-earth is not a matter of economic, scientific or technological expertise, but of practical and ethical wisdom. If Middle-earth had a prophet, he was John Ruskin: 'THERE IS NO WEALTH BUT LIFE.'

But this same world, as we begin *The Lord of the Rings*, is under severe threat from those who worship pure power, and are therefore its slaves — the technological and instrumental power embodied in Sauron (after whom the book itself is named, after all), and the epitome of modernism gone mad. Reading this story, one therefore finds oneself reading our own story. That is one reason why so many readers have taken it so to heart. Another is that just as Sauron is vanquished in *The Lord of the Rings* — albeit barely, temporarily, and at great cost — so Tolkien, crucially, offers his readers hope that what is precious and threatened in our world might survive too. (We shall explore this theme more in the last chapter.)

But the modernists are right, in their own twisted way. *The Lord of the Rings* really *is* a text whose predominant available meanings powerfully contradict their own values; and whose popular success, as a sign of widely shared doubts if not repudiation, makes it, from their point of view, all the worse. In the intention of its author an anti-modernist book, attacking industrialism, secularism, and the myth of Progress, *The Lord of the Rings* squarely falls into the traditions of 'romantic ecology' (Jonathan Bate), 'the ecological perspective of comedy' (Don Elgin), and 'romantic protest' (Meredith Veldman). And like the works of other such authors — William Wordsworth, John Ruskin, William Morris — it has acquired powerful new meanings in a postmodern context. When this dimension overlaps with Tolkien's enduring popularity as a story-teller, in the same way as Dickens, Kipling and Hardy, you get some idea of the potential power of his books, and of the critics' arrogance.

The rubric of 'fantasy,' however, still leaves Tolkien in too large and vague a company. I remarked earlier that the 'mythic mode of imagination' marks him out from, and in many cases gives him an advantage over, many other writers. Giving 'mythic' its full cultural and historical due allows us to see this clearly. It is what raises *The Lord of the Rings* above even excellent books which however embody a more purely personal and idiosyncratic mythology, like David Lindsay's *A Voyage to Arcturus* and Mervyn Peake's Gothic *Gormenghast* trilogy, let alone recent fiction like Lindsay Clarke's *The Chymical Wedding* — a kind of literary 'Twin Peaks,' apparently also drawing on a symbolic/mythic imagination, but in fact mostly occult sensationalism teasingly dressed as metaphysics. Ursula Le Guin's *Earthsea* quartet, Robert Holdstock's *Mythago Wood*, John Crowley's *Little, Big*, Marion Zimmer Bradley's *The Mists of Avalon* and some of Gene Wolfe's books come closer, and in some respects surpass Tolkien; but none can finally sustain his breadth or depth of imagination.

## Disney World

Let me draw two more instructive contrasts. The first is with the most commercially successful purveyor of fairy-tales in the world, in the form of the movie. Then there is the tie-in book, the action figures, the home video, the clothes, the shampoo, the fast food. . . . Yes, it's the Walt Disney Company, which merged with the American Broadcasting Corporation in 1995 to become one of the world's biggest multimedia conglomerates, in a billion-dollar industry second only to the aerospace industry as America's top export earner.

But aren't we talking about some of the same things that inspired Tolkien: old European fairy-tales and folk-tales? Yes and no; but mainly no, because Disney uses them to produce something very different: what Ariel Dorfman aptly calls 'industrially produced fiction.' Disney's versions of the Brothers Grimm really convey the American Dream, that epitome of chrome-plated unsustainability which celebrates the very values that are ripping up what little remains of the world — local communities in a still-enchanted nature — that produced the old stories in the first place. As a former employee remarked, Disney 'could make something his own, all right, but that process nearly always robbed the work at hand of its uniqueness, of its soul, if you will. . . . He always came as a conqueror, never as a servant.'

Exploiting to the full the tendency of film as a medium tending to literal-mindedness, Disney's images violently occupy the mind, gradually destroying the child's imaginative ability to visualize for him/herself. Alternating between brutality and sentimentality (which, as Carl Jung once remarked, is merely a superstructure covering brutality), they communicate not real hope or wonder but commercially-driven imitations. The result is indeed universality, but of a very different kind. The amount of deliberate manipulation that goes into a Disney 'product' cannot be overestimated. 'Pocahontas', for example, was based on American supermodels, but became even more stereotypic during development, 'the rationale being that a chief character has to be iconic enough to work as merchandise and be recognizable to the worldwide market.' All Disney films therefore involve massive market research, and are made in consultation with child psychologists and 'tested' on young audiences before release; characters are rendered deliberately juvenile to exploit protective impulses. Then there are the lucrative spin-offs such as DVDs, merchandise, and tie-ups with (naturally) McDonald's.

Compare this kind of universality to that of Tolkien. His was unforeseen, largely unsought, and very much despite rather than because of the well-named 'cultural industry.' Once again, it is the difference between magic and enchantment. Disney's magic is put at the service of the empire of cultural capital, whose limits of brutality and banality in search of the lowest but most lucrative common denominator have not yet been plumbed. Of course, it contains worse things than 'Snow White': apart from endless glamorized killings and car chases, the banality of 'Reality TV,' MTV and advertising (it's often hard to tell the difference) — and more frightening: the memory-chip 'Soul-catcher' I have already mentioned, for example, or the 'Doppelgänger.' This computer system is currently being developed by MIT's Media Lab. It uses sensors to watch the TV viewer, accumulate information about his/her responses, and alter programming accordingly. Specializing in smile detection, the Doppelgänger thus only shows you images that will make you smile. It is 75% funded by business. 'What we like best,' said a frank spokesman, 'is for our sponsors to make lots of money from our patents . . . and then come back and invest some more.' Small wonder, then, that *everything* soon becomes a commodity in the market-place, even irony about it, just as everyone becomes first and foremost a consumer, and a citizen later, or never.

You can't blame all this on Disney. But it is a major player and founding father of the contemporary cultural industry, which in turn is a billion-dollar part of the contemporary international consumerism. Its fairy-tales thus lie at the very heart of modernity. No wonder that Tolkien once confessed to 'a heartfelt loathing' for all the works of the Disney studios.

Walt Disney was indeed, in his way, a genius. As a matter of historical record, he was also a ruthless, anti-Semitic, sexist, union-busting employer who worked as a Special Agent for J. Edgar Hoover's FBI, cooperating in the witch-hunts of 'unAmerican subversives' by spying on friends and colleagues. But I am not trying to portray Tolkien as some kind of left-wing saint by comparison; and to prove it, my next contrasting example is the late Angela Carter.

## Angela Carter

Angela Carter was a hard-living, independently-minded and strongly feminist writer who produced some of the first and best 'revisionist' fairy-tales for girls and women. She was a complex and controversial writer, and I am only going to touch on the apparent overlap of her work with that of Tolkien. The contrasts are clear enough: in addition to Carter's earthy and sensual feminism, there was a big generation gap between her 1960s anti-authoritarian populism and Tolkien's residually Edwardian love of a quiet, green world. But the two authors drew upon many of the same European and English folk- and fairy-tales; and Alison Lurie thinks Carter shares a Northern air with Karen Blixen, one that I have already said links the latter with Tolkien.

It seems to me, however, that ultimately their projects were exactly opposite. Carter was primarily interested in *dis*enchanting her readers — freeing them from a false glamour cast by a sexist and racist capitalism — whereas he, despite sharing to a surprising extent the same concerns, was trying to work an alternative *re*-enchantment. These represent very different strategies. Neither is necessarily more effective than the other; Carter's sophisticated and anti-mythic subversion of enchantment limits her audience in one way, just as Tolkien's contrary approach does his in another. In terms of appeal discernible through sheer numbers of readers, of course, Tolkien obviously has the edge. But I would also reject the suggestion that, Carter's left-of-centre affiliations notwithstanding, her work is inherently more 'radical.' Indeed, if I am right about the destructiveness of unchecked modernity, then Tolkien's is more needed, and, ironically, less naïve.

Carter's best fiction centres on the circus and the theatre, both arenas whose magic, while potent, falls well within the humanist and secular ambit of drama. This is an art-form which, if Tolkien was right, is necessarily anthropocentric. He contrasted it in this respect with literature, which can admit nature in its own right, which then in turn can enter into art. And literature that hearkens back to ancient myth has a special impetus and ability — perhaps even responsibility — to let the voices of non-human nature speak.

It is also significant that both Carter and Salman Rushdie — the former rightly praised by the latter as 'a thumber of noses, a defiler of sacred cows' — have declared their devotion to *The Wizard of Oz*. For the fundamental point about the Wizard of Oz himself, 'Oz the Great and Terrible,' is this: he was a cheat and a fraud. As such, the tale is a comforting anti-fairy-tale for secular and modernist Grown-Ups; its namesake is just 'a little, old man, with a bald head and a wrinkled face.' '"Hush, my dear . . . don't speak so loud, or you will be overheard — and I should be ruined. I'm supposed to be a Great Wizard." "And aren't you?" . . . "Not a bit of it, my dear; I'm just a common man".'

## Discworld

As a final contrast, let us briefly consider the work of the hugely successful writer Terry Pratchett. His books are stuffed with wizards, witches, trolls, dwarves and magic generally, which would seem to define him squarely as a fantasy writer, albeit a 'comic' one. Yet he is not; at least, not in the same sense as Tolkien. For these are basically literary devices Pratchett uses to produce quintessentially humanist tales: not humanist in the modern sense of a scientistic and universal secular religion — Pratchett's idiom is unmistakably local, that is, English (not even 'British') — but in the best classical, pre-modern tradition (and this is a link with Tolkien) of the human, sceptical and tolerant. It is also comic in Elgin's sense, that is, accepting of nature, the body, and material limits — just what are despised by the intense ambition, individualism and 'spirituality' of tragedy, so beloved of modernity. As such, Pratchett's stories partake of a postmodern emphasis on the local, plural and contingent; they refresh rather than dessicate the contemporary soul.

## Tolkien's True Company

The only books I can think of that seem comparable to *The Lord of the Rings*, in the terms in which I have analysed it here, are other examples of mythic fiction. Those that spring to my mind are Herman Melville's *Moby Dick*, Mikhail Bulgakov's *The Master and Margarita*, Alain-Fournier's *Le Grand Meaulnes*, Russell Hoban's

*Riddley Walker*, and Karen Blixen's autobiographical *Out of Africa*. (There are undoubtedly others, some perhaps among the fantasy novels that I mentioned earlier.) These books too draw their power from a profound and fresh connection with the mythic. And significantly enough, they have also presented literary critics with intractable problems — resulting partly from an absurdly narrow definition of 'literature,' and partly from sheer lack of empathic imagination — and who are thus usually obliged to treat them, at best, as unclassifiable. Except perhaps for Hoban's, they could also be described, like *The Lord of the Rings*, as their authors' life-works.

For all the reasons I have given, this is Tolkien's true company of peers. He is saved by his deep and tough roots in an ancient place from the pathetically deracinated (and therefore shallow) universalism of *Star Wars*, with its bargain-basement Jungian archetypes as eulogized by Joseph Campbell; and by his profound re-creation of myth from the ghastly death-in-life of Disney's commercial imitations, with their plastic grass and 'genuine replica' fairy castles.

> All that is gold does not glitter
>     Not all who wander are lost;
> The old that is strong does not wither,
>     Deep roots are not reached by the frost.

· 6 ·

# CONCLUSION: HOPE
# WITHOUT GUARANTEES

*... the wisdom of the Earth has ways of
re-seeding itself.*

I HAVE USED A PARTICULAR way of approaching and ordering Tolkien's work that he himself suggested when he wrote that
Faërie has three faces: 'the Mystical towards the Supernatural; the
Magical towards Nature; and the Mirror of scorn and pity towards
Man.' In my terms, and reverse order, these are the Shire (humanity), enclosed by Middle-earth (nature), enclosed by the Sea (spirituality). And this schema works well, I think, as far as it goes. But it
is not perfect. First, we must note Tolkien's qualification that of
the three, the essential one is the middle, magical — or rather, enchanted — face turned towards nature. (And let us here assume
that *The Lord of the Rings* qualifies as a modern fairy-story.) Second, I have used the image of the Sea to indicate what imbues but
also transcends the natural world of Middle-earth: the spiritual.
But here my analogy breaks down, because the destructive effects
of the Ring of modernity are not limited to culture and terrestrial
nature; we are now polluting and killing off 'even the everlasting
sea which gave us birth.' Tolkien hints at this too, when at one
point Frodo despairs of his quest: 'Only Elves can escape. Away,

away out of Middle-earth, far away over the Sea. If even that is wide enough to keep the Shadow out.'

In other words, there are at least two reasons — one internal or structural (to do with fairy-stories), and the other external or contingent (to do with the situation of the Earth at this time) — to regard the natural world of Middle-earth as central to Tolkien's tale in terms of what it has to tell us. Another might be that Middle-earth itself is what most of his readers find most memorable, but that is a speculation too far. As an indirect confirmation, however, consider 'The Tale of Aragorn and Arwen.' Despite its poignancy, and its closeness to Tolkien's heart as well as to what he himself considered the essence of his book, this story finds no place in *The Lord of the Rings* proper. Instead, it is consigned to an appendix. The reason may well be that it chiefly concerns death and immortality, and thus belongs with the third sphere or 'face' of the spiritual.

Yet we cannot simply drop the Shire and the Sea, the social and spiritual dimensions; rather, they must be integrated into nature's centrality. Their final synthesis, I think, is that Tolkien's work urges *a new ethic* of human conviviality, respect for life, and ultimate humility. That ethic is to be based on the experience of life on Earth, and therefore the lineaments of life — good earth, clean water, fresh air and the like — as sacred. Finally, for that resacralization to succeed it must be deeply rooted in culture, through being celebrated and communicated in local and traditional ways. The result is not simply a negative critique of 'positivist, mechanist, urbanized, and rationalist culture' but a positive vision of what one reader well described as 'sanity and sanctity.'

This is the heart of the three domains, and thus the chief contemporary 'message' and meaning of Tolkien's work. I do not claim that it is the only possible one, of course. Great art is inexhaustible, mythic art all the more so, and my interpretation inevitably remains, to some extent, personal. But I don't think it is merely personal; this message is real, vital and urgent. What follows is simply how I would flesh it out a bit more.

## *The Elements*

I have already discussed the potent sense of place permeating the natural world of Middle-earth. *The Lord of the Rings* is a work in which an appreciation of this world, culturally mediated but nonetheless sensual, is interfused with an equally powerful sense of its ineffability. In fact, there is a movement within its story from the one toward the other. In the early chapters, and even earlier, in *The Hobbit*, we encounter the hobbits' appreciation of good food and drink, the art of cooking, gardening, bright colours, 'things that make us laugh,' and walking in the countryside. (And that means walking quietly, allowing it to be present — something hobbits were especially good at; not trail-bikes, power-boats and other forms of human domination through machines.)

Gradually, however, the hobbits undergo a transformation, an 'inner' change that is the result of following their quest in the 'outer' world: the hard and dangerous work of destroying the Ring and, for the others, supporting Frodo's effort to do so. The simple sensual appreciation of Pippin's bath song, for example — 'O! Water Hot is a noble thing!' — while undeniably true, gives way to a deeper and truer appreciation of these things, whose aesthetic, spiritual and (literally) vital dimensions are indissolubly one. You might say that without entirely losing their native hobbitish simplicity, they come to an Elvish realization of the nature and value of the essentials of the natural world.

Those essentials sometimes explicitly appear within *The Lord of the Rings*, such as in Pippin's last thought when facing death outside the Black Gate: 'I wish I could see cool sunlight and green grass again!' But they especially arise as Frodo and Sam, in the journey to the heart of their ultimate denial, which is Mount

Doom, are increasingly deprived of them. In Cirith Ungol, 'Earth, air and water all seem accursed.' And as Sam says deep in Mordor, recalling Galadriel's seemingly fantastic earlier offers, or temptations: 'If only the Lady could see us or hear us, I'd say to her: "Your ladyship, all we want is light and water: just clean water and plain daylight, better than any jewels, begging your pardon".' What haunts his thoughts is 'the memory of water; and every brook or stream or fount that he had ever seen, under green willow-shades or twinkling in the sun . . .' Meanwhile, the growing ravages of the Ring on Frodo are having the reverse effect: 'No taste of food, no feel of water, no sound of wind, no memory of tree or grass or flower, no image of moon or star are left to me. I am naked in the dark, Sam, and there is no veil between me and the wheel of fire.' (So much for the supposed other-worldliness and tweeness of *The Lord of the Rings*.)

This kind of perception is nowhere better described than in the nineteenth-century English writer Richard Jefferies' numinous accounts of the elements. He celebrated the earth's firmness that bore him up, the 'pureness, which is its beauty,' of the wandering air, the sea and its strength, and the sun, 'his light and brilliance.' But Jefferies, and this is crucial, was also aware of what he called their soul equivalents. Thus, he wrote of water and sunlight that 'Beside the physical water and physical light I had received from them their beauty; they had communicated to me this silent mystery.' 'Drinking the lucid water, clear as light itself in solution, I absorbed the beauty and purity of it. . . . I look at the sunshine and feel that there is no contracted order: there is divine chaos, and, in it, limitless hope and possibilities.'

The power of such things is such that I think Northrop Frye is right: 'Earth, air, water, and fire are still the four elements of imaginative experience, and always will be.' But here we are speaking of enchantment; the elements of the periodic table, their modern so-called successors, have long since surrendered to what Tolkien called 'the laborious, scientific magician.' That is the key difference between the two. The scientific elements are entirely abstract, in that sense universal, manipulable, and therefore dead; whereas, ironically, the symbolic elements of the imagination require the immediacy of living, sensual, situated and embodied experience in order to enliven and move us.

## *Place*

Such experience is therefore inseparable from a particular place and moment. The place is indeed this Earth, but not in any abstract sense, not even that of 'Gaia.' And it is not just anywhere on the Earth; only actual places on it. In cultures that value their true home, certain of these places become communal sacred sites. As Tim Robinson observes:

> what is hidden from us is not something rare and occult, or even augustly sacred, but, too often, the Earth we stand on. I present to you a new word: 'Geophany.' A theophany is the showing forth, the manifestation, of God, or of a god; geophany is the showing forth of the Earth.

The times of such experiences are not just any times either, but also particular ones, qualitatively charged with meaning, in the ongoing narratives of our lives and of life; stories, as they become. When they become the collective social focus of a culture that has not forgotten its true home, these moments become sacred, tied to the Earth's relationships (that also make life possible) with the Sun, Moon and other cosmic bodies. But those moments, not as abstractions but as experiences, are also inseparable not just from the Earth as a whole but from the places where they happen; so time again brings us back to place.

So too does story. The stories that are oldest and still truest — though they may and must take new forms, like *The Lord of the Rings* — are the farthest from the abstractly universal 'laws' of modern science; they are, rather, what we often patronize and even scorn as myth, folk- and fairy-tale. And myth, of course, is the bedrock of Tolkien's work. Sean Kane, once again hitting the mark, defines it thus:

> Wisdom about nature, that wisdom heard and told in animated pattern, that pattern rendered in such a way as to preserve a place whole and sacred, safe from human meddling: these are the concepts with which to begin an exploration of myth. *Of these, the notion of the sanctity of place is vital.* It anchors the other concepts. . . . Once the power of the place is lost to memory, myth is uprooted; knowledge of the earth's processes becomes a different kind of knowledge, manipulated and applied by man.

Similarly, Gary Snyder has long stressed that 'Of all the memberships we identify ourselves by (racial, ethnic, sexual, national, class, age, religious, occupational), the one that is most forgotten, and that has the greatest potential for healing, is place. . . . Community values (which include the value of nonhuman neighbors in the "hood") come from deliberately, knowledgeably, and affectionately "living in place".'

Conversely, look at the amount of effort in the Bosnian war to destroy not just peoples but their 'tombstones and churches, libraries, theatres and monuments.' Slavenka Drakulic points out that 'these people are not only killing but destroying each other, each other's past, culture, tradition, memory. They are turning this country into a wasteland, as if no one is going to live there any longer, not even a winner . . .' For many people, especially those in whose lives it was a personal lodestar of meaning, the destruction of the Old Bridge in Mostar, built in 1566, was the epitome of this psychotic vandalism. It collapsed on 9 November 1993, at 10.30 am, after repeated shelling. People told Drakulic that 'the bridge is us.' One person cried when she heard about it who had not cried since the war started.

This is a dramatically negative example of the importance of place. But look: the same thing is happening now on a more gradual and piecemeal but hugely greater scale, and its effects will not be very different in the not-so-long run: a global commercial 'pop' monoculture, turning everywhere it touches into the same place, with every aspect virtually interchangeable. As Sue Clifford and Angela King say:

> The bigger the scale the more reduced the sensitivity and the easier it becomes to steamroller strategies for the 'greater good' which prescribe the same solutions to subtly different circumstances encouraging convergence and homogeneity . . . thereby missing the whole point.
> . . . The forces of homogenization rob us of visible and invisible things which have meaning to us, they devalue our longitudinal wisdom and erase the fragments from which to piece together the stories of nature and history through which our humanity is fed. They stunt our sensibilities and starve our imagination.

What we need to recognize are the bonds linking nature, identity and place, whereby the heterogeneity of local distinctiveness 'suggests richness: historical, cultural and ecological.'

So it is true that the vision at the heart of *The Lord of the Rings* is one of the cosmos as 'an organism at once *real, living* and *sacred*'; but this remains too general. Middle-earth's places — each wood and indeed glade within it, streams no less than mighty rivers and individual mountains as much as their ranges, let alone villages, towns and cities: each one unique, and all named not arbitrarily, but as they are natured — these are among its chief glories, and embody its wisdom. When we love our places as these are loved, and as Tolkien encourages us to, then 'longitudinal wisdom' or 'wisdom about nature' can become ours. And that *includes* human nature; for we are a part of it, and there is nothing essential about what it means to be human that is not a refinement and expression, however subtle and unique, of the nature that we share with all other living beings. In this way, difference, uncoerced, finds a common home.

Sauron's desire is just the reverse: to turn everywhere into one empire, ruled by one logic in accordance with one Will. The result of this apparent unity, which can only 'succeed' by being brutally enforced, would be utter fragmentation and isolation, a barely suppressed war of all against all. Note that the symbolic, non-allegorical nature of the Ring is especially important here; it is the wilful exercise of power applied instrumentally to the realization of a single overarching goal. The precise nature of that power — whether primarily economic, religious, political, or whatever mixture of these — is entirely secondary to its intended monism, universalism and homogeneity. The effects of such an enterprise, regardless of the intentions of those who carry it out, are necessarily evil.

In this context, it is also worth recalling that Tolkien more than once asserted that the 'kernel' of his mythology was a place: 'a small woodland glade filled with hemlocks at Roos in York-shire,' where he saw his young wife dancing. This gave creative rise to the 'Tale of Lúthien Tinúviel and Beren,' the heart of *The Silmarillion*, and thence *The Lord of the Rings*.

## Wonder

A resacralization of living nature, and living in nature, is of course a matter of re-enchantment. Its hallmark is the kind of non-utilitarian, open-ended response that I have called, after Tolkien, wonder: as in, 'the realization, independent of the conceiving mind, of imagined wonder.' C. N. Manlove is right that 'there is a very definite and constant character to fantasy, and in nothing is it perhaps so markedly constant as its devotion to wonder at created things, and its profound sense that that wonder is above almost everything else a spiritual good not to be lost.' A fantasy like *The Lord of the Rings* can help us not only to imagine wonder, but realize it by returning to our world and seeing it afresh. It offers renewal not through escapism, but reconnection.

There is a strong relationship with the issue of place here, because without such re-enchantment — that is, unless our response to our communal, natural and sacred places as home partakes of wonder — then the outlook for our places, and all that hangs on them, is very dark. No strategy based purely (or even largely) on calculations of usefulness, self-interest or rationality can survive the onslaught of economic and scientific monologic that comprises 'development.' Only re-enchantment can make it possible to realize that this world, its places and its inhabitants are existentially already wondrous — and as such, worthy of the kind of respect and love that doesn't permit their wanton, callous and stupid destruction. You won't fight for what you don't love.

Another way of understanding enchantment, by the way, is to realize its literal meaning, as voiced by Sam in Lothlórien: 'I feel as if I was *inside* a song, if you take my meaning.' This is not as precious as it might sound, when one recalls the ancient and not-entirely-forgotten sacred and ritual aspects of song. Their Paleolithic origins are still preserved in the Aboriginal 'songlines,' embodied in a sacred landscape. Wisdom of the Earth, place, and enchantment thus once again coincide.

But there is a problem with re-enchantment as a program; or rather, two related problems. The first is that it cannot be straightforwardly adopted and pursued *as* a deliberate and planned program, because that simply re-admits humanist utilitarianism by the back door. In Tolkien's terms, it would be the application of a technique, rather than an art; and as such, not enchantment at all, but magic. And you cannot use the Ring against the Ring without becoming the Enemy. So re-enchantment can only encourage, amplify and build on instances where it is already happening spontaneously, as it were. It must be a 'collective spirituality,' not a new religion.

The second problem is the vulnerability of enchantment to

magic. Just as the One Ring of Power potentially dominates the three Elven rings — and actually, once Sauron is directly in charge of it — there is evidently no form of enchantment, no matter how personal or elusive (natural, sexual, artistic, even political), that cannot be exploited and corrupted by the magic of capital/the state/techno-science. The power of the Ring is not the only kind, to be sure; but it is the 'greatest in Middle-earth.' Yet I have also argued that re-enchantment is urgently needed. Is there any way out of this dilemma? What does Tolkien's work suggest?

## Hope

In *The Lord of the Rings*, when the One Ring was destroyed, the power of the Three faded too, albeit only slowly. This heralded the onset of what feels like our own disenchanted age: 'the time comes of the Dominion of Men,' to quote Gandalf, 'and the Elder Kindred shall fade or depart.' But I take our situation to be (perhaps perpetually) rather that of the book's beginning, when the Ring is still abroad and unvanquished; and the unfolding of the tale itself to be Tolkien's best hope for a happy 'ending.' In that case, re-enchantment in defence of place is still vital. But given its vulnerability, there is only hope that it will be able to act if we collectively renounce, and thus destroy, the Ring of Power. And the prospects of this happening are unclear. On the one hand, of course, even Frodo fails the final test. On the other, the Ring *was* destroyed.

I think Tolkien saw wholly voluntary renunciation of Power as utopian. The circumstances of the Ring's final destruction suggest, I'm afraid, that such a thing can only come about in the context of a crisis, some kind of collapse of what power has wrought, that is sufficiently serious to reveal the modern magic of the One Ring in its true colours, stripped (above all) of its present glamour. This would provide an opportunity to realize just how inflated are its claims, how dangerous its means, and how destructive its goals.

Such a crisis could take various forms, including a general failure of the machines (especially computerized) into which modernity has put so much of its power. It might involve an economic crash, ecological disaster, and/or general political failure. But it would have to be pretty widespread. At the same time, if it is so severe as to threaten even recovery, then that too would be hopeless. I don't mean a return to our current 'normal' conditions and demands, obviously, but recovery as Tolkien defines it — the 'regaining of a clear view . . . so that the things seen clearly may be freed from the drab blur of triteness or familiarity — from possessiveness' — through a rekindling of the wonder of the natural world, including humanity when not Ring-driven.

If Tolkien is right, a renunciation of the Ring could come about like this; but such an outcome is by no means assured, and we would still need 'luck' and whatever our accumulated reserves may be of Pity and Mercy. And as we have already seen, he openly denies that it would be permanent: 'Always after a defeat and a respite, the Shadow takes another shape and grows again.' Here, however, we must be careful not to become infected by Tolkien's own pessimism (as it seems to me). Faced with disappearing communities, nature, and spiritual values, he tried to salvage and represent them in a form itself held increasingly in contempt, that of mythic story. Indeed, his own personal epitaph might be the parting words to Aragorn or Estel (Elvish for hope) from his mother: 'I gave Hope to the Dúnedain' (the human races from the West) 'I have kept no hope for myself.'

But what Tolkien called 'Hope without guarantees' is vital. It permits wonder to act, and this in turn enables places to be defended. Then this mutually nourishing trinity, in which each helps the other to survive and thrive, can increasingly overtake its mirror-image cycle: hopelessness, nihilism, and nowhere, which also feed off each other. So no matter how dark the future appears we must, like Frodo, Sam and (just as importantly) Gandalf, refuse despair. Even more importantly, there *is* still hope. The future is not fixed. There are still places on Earth that are beautiful, and loved, and cared for; there are still wonders, people who can wonder, and indeed who work wonders. And that is what *The Lord of the Rings* itself, which is what people read and what therefore chiefly matters, conveys: Arda (the Earth) unmarred, and even healed. That includes us.

It is also significant that Tolkien's story does not depict a single, all-consuming crisis; the War of the Ring dominates, but not absolutely everything, and not forever. Nor is the end of the Ring a purely voluntary, willed and idealistic renunciation. Instead, it is made possible by countless acts of courage, kindness and help, both small and great, from unknown people and forces, in unforeseen circumstances, that together provide an opportunity to do the right thing. It is preceded and succeeded by a commitment to the simple good things of life — food, water, green and growing things — that extends through conviviality and creativity to an appreciation of life itself, at once natural and spiritual, as the ultimate value. And it succeeds as much, if not more, by the efforts of the humble and ordinary as those of the mighty.

This realization is what, I believe, Tolkien's readers find, above all else, in his books. 'It is a fair tale, though it is sad, as are all the tales of Middle-earth, and yet it may lift up your hearts.' Tolkien's Middle-earth gleams with the light of an ancient hope: peace between peoples, and with nature, and before the unknown.

# AFTERWORD

SEVEN YEARS HAVE passed since I wrote *Defending Middle-Earth*. Re-reading it now, there is nothing important I would change or retract, but I welcome the opportunity to add a few points. Above all, I welcome the chance for North American readers to read it and make up their own minds about what it says.

Recently, of course, the movies have brought *The Lord of the Rings* many new readers; together with *The Hobbit*, it sold 24 million copies in 2001–02 in the United States alone. But this is only the latest wave of a process that began long before. For forty of its fifty years of existence, and throughout the English-speaking world, Tolkien's masterpiece has always managed to win over new generations, even when it was unfashionable. It is worth asking once again: why?

The power of Peter Jackson's interpretation of Tolkien as visual and even moral drama is undeniable. Perhaps its chief failing — not what was left out, but changes for the worse in the story's plot and characters — only serves to emphasize Tolkien's own narrative skills. But I doubt that even the best storytelling, although it meets a deep public hunger, is alone sufficient to explain his popularity. Why this particular book, with its extremely unlikely story, cast and setting?

That is what I have tried to address in this little work. Contrary to some suggestions, I nowhere argue that Tolkien was himself a postmodernist, nor that ecology is the only or even most important key to his work. Nor have I ignored his warnings against reading his stories allegorically. My subject is the *applicability* of his work in the experience of his readers (which he defended). And what I do insist is that in order to understand its extraordinary appeal, we must look at its content, and its meanings to contemporary readers. As Tolkien himself wrote, after everything that research into origins can discover, 'there remains still a point too often forgotten: that is the effect produced *now* by these old things in the stories as they are.'

That appeal has two faces. Positively, *The Lord of the Rings* imaginatively re-connects its unprejudiced readers with a world that is still enchanted, that is, a world in which nature — including, but also greatly exceeding, humanity — is still mysterious, intelligent, inexhaustible, ensouled. Its places each have their own personality, and its spirits each have their home. (What a relief for human beings to no longer feel they have to shoulder the burden of sentience and agency alone, in an otherwise dead and meaningless universe!) Middle-earth is thus a strange world to most of us, but it is also recognizable as one we used to live in but since have lost . . . although not, it seems, altogether forgotten. And that connection opens the door to the possibility of realizing that our own world (once seen aright) is *still* enchanted.

Tolkien's own definition of enchantment, by the way — 'the realization, independent of the conceiving mind, of imagined wonder' — also remains one of the best. One synonym could be, 'sacred.' Another might be, 'NOT FOR SALE.'

The negative aspect of Tolkien's appeal is the seriousness and scale of the threat to such an enchanted world in Middle-earth at the end of the Third Age, whereby readers can easily (and without being lectured) recognise how imperilled is our own. As I argue, the threat is principally to three literally vital values: (1) community, including, but not limited to, the family (the hobbits);

(2) the nonhuman natural world (Middle-earth itself); and (3) that dimension of life which cannot be quantified, controlled or exploited which we call 'spiritual.' And in all three spheres, the crisis which I described seven years ago has continued to accelerate. We are living in a world even more at the mercy of imperial fantasies of Progress. The words with which Saruman tempted Gandalf — 'Knowledge, Rule, Order' — are still the velvet glove of the iron fist: one rule, by one power; and when they fail to persuade, we are told it is all 'inevitable' and 'unavoidable' in any case, and those who continue to resist are treated as traitors. (Tolkien's essay 'On Fairy-Stories' remains unsurpassed on this point, among others.)

Of course, there are very different versions of this ambition. But what marks them all out as such — whether economic, political, religious or cultural — is the claim of singular and universal truth, and the aspiration to complete control. And only one hand, as Gandalf pointed out to Saruman, can wear the One Ring.

By the same token, we stand in more need than ever of viable alternatives — starting with the very idea of alternatives. And those necessarily begin with a *vision* of alternative futures that defy the attempt to corral us all into the iron cage of modernity (to use Max Weber's prescient image). *The Lord of the Rings* offers us one such vision of an alternative world, where enchantment — communal, natural, and spiritual — survives (albeit barely) the onslaught of modernity at its most virulently pathological. And it comes in the form we most naturally respond to: a story. Is it any wonder, then, that so many contemporary readers keep turning to Tolkien, and finding there hope and renewal?

More than ever, we need reminding of the courage that lies buried deep in the hearts of unsung 'small,' ordinary people; of the wonder of what we have too long taken for granted: living water, fresh air, clean earth; of something as basic as good food cooked with care and eaten with appreciation in convivial company (compare the oxymoron of a 'power lunch'); of our souls' hunger for such 'useless' activities as (say) walking in woods under starlight; and of the value of loyalty, honour and friendship which, like all such things, cannot be measured, weighed or expressed in dollars and cents. And these days, we need reminding of the truth of what Gandalf and Aragorn realized: that military force is, at best, a stop-gap measure to buy time. The real work, slow and un-glamorous, grows out of compassion (literally, feeling together with) and co-creating a better world.

Such applicability confirms Tolkien's realism and continu-ing relevance. We should also remember his warning against lit-eral-minded allegory, however. There are also certainly epicenters of malevolence, but no one single Barad-dûr. And servants of Orthanc — mindlessly managerial, religiously rationalist and, of course, masters of double-speak ('War is peace') — seem to be everywhere. Conversely, there are hobbits and heroes in the Mid-dle East as well as America, and in Africa as much as Europe. Actually, anyone defending the values at the heart of Tolkien's work (community, nature and the spirit) is an honorary member of the Fellowship of the Ring.

✌

As my title implies, this book places great emphasis (although by no means exclusively so) on nature, and on Tolkien's defence of it. What that defence means to his readers cannot be separated from their own relationship with nature, and the overall current context for that in turn is human-driven mass extinction: in a word, ecocide. Since this Earth is (except in the mad dreams of New Age gnostics and Space Age scientoids) our only home, that amounts, of course, to suicide. But Tolkien's concern rightly ex-tends beyond anthropocentrism to our savage destruction of count-less other life-forms and the places that are their homes as well.

I have been accused of using Tolkien to advance an ecological agenda. But nothing in this book about defending nature does not draw its warrant from the contents of Tolkien's own work; and I would add that although it is not provable or strictly relevant, I believe he himself would have thoroughly approved. We are talking, after all, about a man who (in his letters) identified 'the servants of the Machines' as a powerful new privileged class, denounced 'the utter folly of these lunatic physicists' who created the atomic bomb, and described as the 'one bright spot' of the present world 'the growing habit of disgruntled men of dynamiting factories and power-stations'!

Of course, trees were Tolkien's special concern ('In all my works I take the part of trees as against all their enemies'). And everything that has happened since his books first appeared has borne out his fears. The destruction of trees, both outright and their conversion into the green deserts of industrial forestry, continues to intensify, not only in places like Brazil and Indonesia but closer to home: *five percent* of native forest now remains in the continental United States. And it is, to quote the title of a recent book on the global assault on forests by Derrick Jensen and George Draffan, *Strangely Like a War*. Not for nothing, in this respect as in others, is Tolkien's book an account of the War of the Ring.

It has also been suggested (by Verlyn Flieger) that Tolkien was confused, or at least inconsistent, on this subject; that from nature's point of view, there is no difference between, say, the hobbits of Bucklebury cutting back the Old Forest and Saruman turning Fangorn into fuel for his war-furnaces. Flieger also thinks Tolkien shrank from recognising that civilization is necessarily locked into a war with nature. But this is a misunderstanding in a number of ways. Most obviously, as that example shows, it oddly fails to distinguish limited self-defence (the human right to which, when it is necessary, I do not deny) from gross exploitation finally resulting in complete destruction.

Secondly, Tolkien vigorously asserted, as I argue, just the opposite: that civilization can, and must, learn to co-exist with nature. (It doesn't take any great insight to see that any 'victory' by the former would ultimately be self-defeating.) He was contesting a point — in this case, a fatalistic cult of modernism — not ducking it.

Finally, there is a confusion of wilderness with nature. Wilderness is immensely important but it does not exhaust the natural world. It was *wildness*, not wilderness, that Thoreau suggested contained the salvation of the world. Nature should be understood (in the words of the philosopher David Wiggins) 'not as that which is free of all trace of our interventions . . but as that which has not been entirely instrumentalized by human artifice, and as something to be cherished . . . in ways that outrun all considerations of profit.' So an absence of 'pure' wilderness does not mean there is only culture; these are both unsupportable extremes.

Industrial forestry, for example, is not the only alternative to untouched primary forest. Much of Tolkien's attitude to trees reflects a quite different and much older perspective, namely, *woodsmanship:* a sensitive and sustainable use of nature, not for profit but for life, which entails not the conquest of an objectified nature but an ongoing relationship with various subjectivities, many of them nonhuman. There will be conflicts, of course, just as there are among humans. But the ultimate sense — which is obvious in all of Tolkien's work — is of a world that is *shared;* and far from confused, that insight is profoundly realistic. (By the way, for those interested in further reading, the best guides I know to such a view/world are David Abram's *The Spell of the Sensuous* [1996] and Sean Kane's *Wisdom of the Mythtellers* [1998].)

Returning to the movies for a moment, Tolkien himself believed that because 'the characters, and even the scenes, are in Drama not imagined but actually beheld — Drama is . . . an art fundamentally different from narrative art. Thus, if you prefer Drama to Literature . . . , [y]ou are, for instance, likely to prefer characters, even the basest and dullest, to things. Very little about trees as trees can be got into a play.' Personally, I agree. Despite the dramatic New Zealand setting (and Jackson's best efforts with the Ents), that is just the problem: its drama was ultimately only a setting for the more-or-less human characters. There was very little sense of something essential that permeates the entire book: Middle-earth itself, and almost all its places, as living, intelligent personalities. (And without a vivid and profound grasp of just that, no ecological or environmental program stands much chance of success.)

～

Still, I now think I did overplay the place of nature in Tolkien's work in a different way, one that follows from my own three-fold schema. For just as the natural world of Middle-earth contains the Shire, the world of the hobbits, so its mountains, forest and rivers are enclosed by the encircling Sea of the spirit.

Now I would like to stress that these spheres do not replace each other. Quite the reverse: the natural world nurtures and supports human society and culture, and without such support they would quickly vanish. By the same logic, however, Tolkien was suggesting that nature too is sustained, and even dependent upon, the spiritual, without which living more-than-human nature dies and is replaced by its corpse: the inert, quantified and commodified object we are now taught to perceive. In this sense, then, sacrality has the last word . . . or enchantment, to use Tolkien's preferred term: the experience, so to speak, of the spiritual.

We should be careful about this. The spiritual escapes the kind of easy and precise definition which is itself a kind of exercise in disenchantment. A couple of things should be said here, however. One is to repeat the crucial distinction between the spiritual and religion. As Max Weber saw long ago, religion itself becomes an enemy of enchantment when it asserts its own sole universal truth, and thus becomes entangled in aspirations to complete control and ultimate power. In contrast, spirituality — following the nested relationship of the Sea, Middle-earth and Shire — is embodied in nature, including our bodies, and embedded in our shared social lives. So it is necessarily ongoing and plural, with as many forms and perspectives as there are of life itself. And in our Middle-earth, as in Tolkien's, those are by no means limited to humans, or even to biological entities!

That brings me to the second point. In reading *The Lord of the Rings*, we can imaginatively experience a world where, for example, even certain places (particular mountains, forests, rivers) are alive — as they commonly used to be, until we started to see them otherwise. Sanity and sanctity are everyone's common birthright, but learning (and/or unlearning) our way back to them — without which they do not become *real* — requires a personal journey that no-one else can can take for you: '. . . the grey rain-curtain turned all to silver glass and was rolled back, and he beheld white shores and beyond them a far green country . . .' Tolkien's right renown should be for making that trip possible to imagine, and easier to undertake, for many millions of people. So his defence of Middle-earth is fully as spiritual as it is ecological and cultural. But it is not a journey away from our lives and our home here on Earth; ultimately, and crucially, it is a *return*. The last words of the whole book, after all, are Sam's: 'Well, I'm back.'

Odds and ends: the paper I half-seriously mentioned in my Preface did itself eventually find a home, thanks to Walking Tree Publishers, based in Zurich and Berne. Entitled "Tolkien and the Critics: A Critique", it can be found in a collection called *Root and Branch: Approaches Towards Understanding Tolkien*, edited by Thomas Honegger (1999). Although it has dated in some ways (I wouldn't be quite so free with 'postmodern' now; 'hypermodern' might be better), the need for it has not. Here in the United Kingdom, the BBC recently polled three-quarters of a million readers to discover the nation's favourite book. The final show included a panel of three literary professionals who were asked to comment on each of the top ten choices, and despite the fact that *The Lord of the Rings* led throughout the program's eight weeks (and won), two out of three hadn't bothered to read it. I would be surprised if the situation was very different in the United States.

Finally, I have a few more people to thank, some for the second time. I have received long-standing support, which I deeply appreciate, from Tom Shippey (who single-handedly redeems the good name of Professors of English); John Garth (whose brilliant biography of the first part of Tolkien's life was itself a struggle against the odds); Charles Noad (for his knowledge and love of Middle-earth, plus a shared appreciation of a proper pint of 1420); David Doughan (for cultured camaraderie, always in a good cause); David Abram (for first suggesting a North American paperback, and igniting my own enthusiasm for it); and Sheryl Cotleur (for 'getting it,' fully and unconditionally, and encouraging me to share it more widely). My thanks also go to Clay Harper for making this edition a reality. May his wager bear fruit!

⤜⤝

Patrick Curry can be contacted at dme@gn.apc.org

# References

*The number is the page number on which the relevant passage ends.*

## 1. Introduction

1.

Epigraph: Abraham Lincoln. (Something very similar is also attributed to Max Beerbohm.)

J.R.R. Tolkien, *The Lord of the Rings* (London: Grafton Books, 1991), I, 66. (Henceforth *LoR*; all volume and page numbers given are from this edition.)

2.

Sales: based on figures supplied to me by HarperCollins and by Houghton Mifflin (courtesy Richard McAdoo), and on those in John Ezard, 'Tolkien's Shire,' *The Guardian* (28–29.12.91). According to *The Guinness Book of Records* (1993), Jacqueline Susann's *Valley of the Dolls* (1966) and Margaret Mitchell's *Gone with the Wind* (1936) have sold about 28.5 and 27 million copies respectively. (No mention is made of *The Hobbit*.)

Translations: based on information kindly supplied by HarperCollins in 1992.

Vietnamese: *The Sunday Times* (6.8.1967) (with thanks to Charles Noad).

Tolkien unfashionable: see James Park, *Cultural Icons* (Bloomsbury, 1991).

Present popularity: in a recent survey of readers' 'favourite novel' by *The Sunday Times* (24.9.95), with almost 1100 respondents, *The Lord of the Rings* came second (behind *Pride and Prejudice*). A survey of teenagers' reading habits showed *The Lord of the Rings* still high among 15–16 year-olds (*The Guardian*, 16.12.95). In England, at least, my assertion can be confirmed by talking to the relevant buyer for virtually any bookstore. On the HarperCollins takeover, see Charles E. Noad, 'Announcements,' *Amon Hen* 105 (Sept. 1990).

Libraries: Public Lending Right figures; see also the *Times Literary Supplement* (14.1.94), and *The Guardian* (7.1.93).

First editions: M. Hime, writing in *Firsts* 5:10 (Oct. 1995) 41. (Thanks to Steele Curry for pointing this out.)

3.

Gollum's Channel: James Hamilton-Paterson, *Seven-Tenths: The Sea and Its Thresholds* (London: Hutchinson, 1992), 35.
Big money: except, of course, treasure in *The Hobbit*.

5.

Critical incomprehension: Although I didn't realize it until after it was written, the entire theoretical thrust of my work on Tolkien is encompassed within Barbara Hernnstein Smith's brilliant *Contingencies of Value: Alternative Perspectives for Critical Theory* (Cambridge MA: Harvard University Press, 1988).

I should add that my basic thesis here regarding Tolkien's critics and readers was confirmed, as if in a laboratory test-tube, by the Waterstones survey of more than 25,000 readers in England, who were asked to select the one hundred most important books of the century. Announced in January 1997, the result — with *The Lord of the Rings* in clear first place — was greeted with yet another display of ignorant critical dismay and contempt from the likes of Auberon Waugh, Germaine Greer (although not Malcolm Bradbury) and columnists like Catherine Bennett. See my 'Commentary,' *New Statesman* (31.1.97).

The best works on Tolkien that I have read (without claiming to be exhaustive) are: T. A. Shippey, *The Road to Middle-Earth* (London: George Allen & Unwin, 1992 [1982]); Brian Attebery, *Strategies of Fantasy* (Bloomington IN: University of Indiana Press, 1992); Don D. Elgin, *The Comedy of the Fantastic: Ecological Perspectives on the Fantasy Novel* (Westport: Greenwood Press, 1985); Ursula K. Le Guin, *The Language of the Night: Essays on Fantasy and Science Fiction*, ed. Susan Wood (London: The Women's Press, 1989); Verlyn Flieger, *Splintered Light: Logos and Language in Tolkien's World* (Grand Rapids: Wm. B. Eerdmans Publ. Co., 1983); and Brian Rosebury, *Tolkien: A Critical Assessment* (London: St Martin's Press, 1992).

6.

Edmund Wilson, 'Oo, Those Awful Orcs!,' *The Nation* 182:15 (14.4.1956); Philip Toynbee, 'Dissension among the Judges,' *The Observer* (6.8.1961).
'Paternalistic,' etc.: Walter Scheps, 'The Fairy-tale Morality of *The Lord of the Rings*,' 43–56 in Jared Lobdell (Ed.), *A Tolkien Compass* (La Salle: Open Court, 1975), 52. See also R. C. West, *Tolkien Criticism: An Annotated Checklist* (Kent: Kent State University Press, 1970; rev. edn., 1981); Judith A. Johnson, *J.R.R. Tolkien: Six Decades of Criticism*, Bibliographies and Indexes in World Literature, No. 6 (Westport CT: Greenwood Press, 1986); Wayne G. Hammond, 'The Critical Response to Tolkien's Fiction,' 226–32 in Patricia Reynolds and Glen H. GoodKnight (Eds.), *Proceedings of the*

*J.R.R. Tolkien Centenary Conference* (Milton Keynes: The Tolkien Society, and Altadena: The Mythopoeic Press, 1995); and Tom Shippey, 'Tolkien as a Post-War Writer,' 84–93 in Reynolds and GoodKnight, *Proceedings*. My paper/book mentioned in the Preface, wherever it eventually finds a home, will take on Rosemary Jackson, Christine Brooke-Rose, Fred Inglis, Jack Zipes and others in more detail than is appropriate here. (When it comes to Tolkien, it's like shooting fish in a barrel, and rather large ones at that; but someone's got to do it.)

Fatuous witticisms: John Goldthwaite, *The Natural History of Make-Believe: A Guide to the Principal Works of Britain, Europe and America* (Oxford: Oxford University Press, 1996); Michael Moorcock, *Wizardry and Wild Romance* (London: Victor Gollancz, 1987), 125.


7.

Interpretations: respectively: Bruno Bettelheim, *The Uses of Enchantment: The Meaning and Importance of Fairy Tales* (Harmondsworth: Penguin, 1978); Angela Carter (Ed.), *The Virago Book of Fairy Tales* (London: Virago, 1990), and Marina Warner, *From the Beast to the Blonde: On Fairy Tales and their Tellers* (London: Chatto and Windus, 1994); Vladimir Propp, *Theory and History of Folklore*, ed. Anatoly Liberman (Manchester: Manchester University Press, 1984); C. G. Jung, 'The Phenomenology of the Spirit in Fairytales,' 207–54 in *The Archetypes and the Collective Unconscious* (Princeton: Princeton University Press, 1968; second edn.), and Marie-Louise von Franz, *Interpretation of Fairy Tales* (Dallas: Spring Publications Inc., 1970); Rudolf Meyer, *The Wisdom of Fairy Tales* (Edinburgh: Floris Books, 1988 [1929]); Jack Zipes, *Breaking the Magic Spell* (London: Heinemann, 1979). (Only the last specifically discusses Tolkien.) For quite an enjoyable Jungian tour of Middle-Earth, see Timothy R. O'Neill, *The Individuated Hobbit* (Boston: Houghton Mifflin, 1979). There is also Robert Bly, *Iron John. A Book about Men* (New York: Addison-Wesley Publ. Co. Inc., 1990); and see a recent booklet by Derek Brewer, *The Interpretation of Fairy Tales: Implications for Literature, History, and Anthropology* (Austin: The University of Texas, 1992).

Blindness: one shining recent exception is Roberto Calasso, *The Marriage of Cadmus and Harmony* (London: Jonathan Cape, 1993); another (although a very different book) is Sean Kane, *Wisdom of the Mythtellers* (Peterborough Ont.: Broadview Press, 1994). One must also exempt Robert Bly, whatever his other offences may be; just as one cannot Marina Warner, whatever her other virtues.

Class alliance: Zipes, *Magic Spell*, 151–2.

Inside/outside: One 'fantasy' writer, John Crowley, has incorporated a variation of this principle into his best work, *Little, Big* (London: Victor Gollancz, 1982): namely, 'the further in you go, the bigger it gets' (p. 43). Later, I remembered encountering the same idea in C. S. Lewis, *The Last Battle* (London: The Bodley Head, 1956), 180: 'The farther up and the farther in you go, the bigger everything gets. The inside is larger than the outside.' But I have only just discovered it in G. K. Chesterton's *Autobiography*, published after his death in 1936: 'in everything that matters, the inside is much larger than the outside.' The best discussion of this whole issue that I have seen is in Owen Barfield's *Speaker's Meaning* (Middletown: Wesleyan University Press, 1967).

Determined and defined place: nor can Tolkien's philology and invented languages on any account be left out; see Shippey, *Road*.

8.

Tolkien on allegory: *LoR*, I, 11–12.

No allegory: Humphrey Carpenter (Ed.), *The Letters of J.R.R. Tolkien* (London: George Allen & Unwin, 1981), 262 (henceforth *Letters*, with page rather than letter numbers given).

He that breaks: *LoR*, I, 339.

Shippey, *Road*, 72.

Max Luthi, *The Fairytale as Art Form and Portrait of Man*, transl. Jon Erikson (Bloomington: Indiana University Press, 1984), 145.

9.

Present-day writers: good examples among the British literati include Salman Rushdie, Kazuo Ishiguro, Margaret Drabble and Martin Amis. The last, as Pullman says, is more concerned to show you he is telling you a story than actually to tell it; cleverness is all. But if Tolkien's work has an opposite, it is surely the deathly quotidian non-narratives of Nicholson Baker.

Tolkien's language and style: tackled chiefly by Shippey, in *Road*. See also Derek S. Brewer, '*The Lord of the Rings* as Romance,' 249–64 in Mary Salu and Robert T. Farrell (Eds.), *J.R.R. Tolkien, Scholar and Storyteller* (Ithaca NY: Cornell University Press, 1979), 261–2; and Rosebury, *Tolkien*.

Capacity for imagination: I owe this point to Virginia Luling, in conversation.

Hugh Brogan, 'Tolkien's Great War,' 351–67 in Gillian Avery and Julia Briggs (Eds.), *Children and Their Books: A Celebration of the Work of Iona and Peter Opie* (Oxford: Clarendon, 1989), 358. See also Shippey, *Road*, and Rosebury, *Tolkien*.

10.

J.R.R. Tolkien, 'On Fairy-Stories,' 9–73 in his *Tree and Leaf* (London: Unwin Hyman, 1988), 32. (This is a very important essay for any consideration of the subject, and/or of Tolkien's own work.)

12.

Modernity: see Stephen Toulmin, *Cosmopolis. The Hidden Agenda of Modernity* (Chicago: University of Chicago Press, 1990); also Zygmunt Bauman, *Intimations of Postmodernity* (Oxford: Basil Blackwell, 1992); and Paul

Ekins, *A New World Order: Grassroots Movements for Global Change* (London, Routledge, 1992). See also William Greider, *One World, Ready or Not: the Manic Logic of Global Capitalism* (New York: Simon and Schuster, 1997).

13.
Bauman, *Intimations*, x–xi. *Cf.* Thomas Doherty, *Postmodernism: A Reader* (Hemel Hempstead: Harvester Wheatsheaf, 1993), 5: 'In the desire to contest any form of animistic enchantment by nature, the Enlightenment set out to think the natural world in an abstract form . . .' See also Paul Feyerabend, *Farewell to Reason* (London: Verso, 1987).

14.
Wisdom: what Aristotle called *phronesis*, and distinguished from *theoria*, or precise and certain knowledge.
John Ruskin, 'The Moral of Landscape,' *Modern Painters* III (1856), 17.36.
Bauman, *Intimations*, 222.

15.
State/atom bomb: *Letters*, 63, 116.
Motive: *Letters*, 418.
Fraser Harrison's excellent essay, 'England, Home and Beauty,' 162–72 in Richard Mabey with Susan Clifford and Angela King (Eds.), *Second Nature* (London: Jonathan Cape, 1984), 170. (It is absurd that this powerful collection has been allowed to go out of print.) *Cf.* Kenneth Burke, *Counter-Statement* (Chicago: University of Chicago Press, 1957 [1953]), 119: 'people have gone on too long with the glib psychoanalytic assumption that an art of 'escape' promotes acquiescence. It may, as easily, assist a reader to clarify his dislike of an environment in which he is placed.'

16.
History and myth: Tolkien, 'Fairy-Stories,' 31.

17.
'Fairy-Stories,' 28. (But Tolkien adds that 'The essential face of Faërie is the middle one, the Magical.')
Local distinctiveness: this phrase comes from a campaign by Common Ground, a unique environmental and cultural charity based in London whose concerns resonate closely with many both of Tolkien and of this book. See Sue Clifford and Angela King (Eds.), *Local Distinctiveness: Place, Particularity and Identity* (London: Common Ground, 1993).
Uncanny feeling: Janet Menzies, 'Middle-Earth and the Adolescent,' 56–71 in Robert Giddings (ed.), *J.R.R. Tolkien: This Far Land* (London: Vision Press, 1983), 57.

18.
Loved places: in the broadest sense of 'people.' I have in mind Lothlórien, for example, but also much of the Shire.

19.
Gregory Bateson, *Steps to an Ecology of Mind* (New York: Ballantine Books, 1972), 462.

20.
Jaded: this is a recent encomium in *Time Out* (5–12 Oct. 1994) by its chief art critic, Sarah Kent, commending an exhibition of realistically mutilated and dismembered shop-window dummies: 'They satisfy your blood-lust, they seduce, and they make you sick. Brilliant.' And where do you go from here?
Ihab Hassan, 'Pluralism in Postmodern Perspective,' 196–207 in Charles Jencks (Ed.), *The Post-Modern Reader* (London: Academy Editions, 1992), 204.
Tolkien, 'Fairy-Stories,' 51. I think he was impelled by something else too: his recognition, practically extinct in the academy and rare outside it, of the power of living myth.
English epic tradition: *Letters*, 231, 144.

21.
E. M. Forster, *Howards End* (London: Penguin, 1989), 262.
The Normans: Robert Bartlett, *The Making of Europe: Conquest, Colonialization and Cultural Change 950–1350* (London: Allen Lane, 1983), 272; Albert C. Baugh and Thomas Cable, *A History of the English Language*, third edn. (London: Routledge, 1978), 113, 116; Lincoln Barnett, *The History of the English Language* (London: Sphere, 1970), 108–9; Peter Berresford Ellis, *Celt and Saxon* (London: Constable, 1993).

22.
Farmer Maggot: *LoR*, I, 182.
Virginia Luling, 'An Anthropologist in Middle-Earth,' in Reynolds and Good-Knight, *Proceedings*, 53–57. For an account of the origins of the Europeanization of Europe, see Bartlett, *Conquest*.

23.
'Heart of a heartless world': Karl Marx, 'On Religion.'

## 2. The Shire: Culture, Society and Politics

24.
Epigraph: Hubert Butler, *Escape from the Anthill* (Mullingar: The Lilliput Press, 1986), 95.
J.R.R. Tolkien, *The Hobbit* (London: Grafton Books, 1991), 15. (Henceforth *Hobbit*; all page numbers are from this edition.) *LoR*, I, 17, 18; *Hobbit*, 16; *LoR*, II, 23. 'Anglo-hobbitic': Shippey, *Road*, 108.

25.
Hobbits: *LoR*, I, 59; III, 173; II, 202; I, 50, 26, 219.
Frodo: *LoR*, I, 25, 67.
Sam: *Letters*, 105, 329 (on Sam).

26.
Shire: *LoR*, I, 28, 29, 75.
Shire/rural England: *Letters*, 250, 230; and see Clyde Kilby, *Tolkien and The Silmarillion* (Berkhamsted: Lion Publishing, 1977), 51. The hobbits' obsession with family genealogy is an Icelandic touch, however.
George Orwell, 'The Lion and the Unicorn: Socialism and the English Genius,' 527–64 in *Collected Essays, Journalism and Letters* (London: Secker and Warburg, 1968).

27.
Warm bath: *Hobbit*, 113.
'Accommodate modernity': Hugh Brogan, 'Great War,' 360. *Cf.* Shippey, *Road*, 65.

28.
'English yeomen': David Harvey, *The Song of Middle-Earth: J.R.R. Tolkien's Themes, Symbols and Myths* (London: George Allen & Unwin, 1985), 114.
Less noise: *Hobbit*, 15.
Martin J. Weiner, *English Culture and the Decline of the Industrial Spirit 1850–1980* (London: Penguin, 1985), 47, 61 (Sturt), 81, and Chapter 4 generally. See also Alun Howkins, 'The Discovery of Rural England,' 62–88 in Robert Coll and Philip Dodd (Eds.), *Englishness. Politics and Culture 1880–1920* (London: Croon Helm, 1986).
Weiner, *English Culture*, 47, 49.

29.
Catherine R. Stimpson, *J.R.R. Tolkien* (New York: Columbia University Press, 1969) (Columbia Essays on Modern Writers No. 41), 8. See Hal Colebatch's critique of Stimpson in his *Return of the Heroes: The Lord of the Rings, Star Wars and Contemporary Culture* (Perth: Australian Institute for Public Policy, 1990), 61–6.
Historical accident: *Letters*, 197.
Land/fatherland: Jonathan Bate, *Romantic Ecology: Wordsworth and the Environmental Tradition* (London: Routledge, 1991), 11.
Stimpson, *Tolkien*, 13.

30.
Orcs: *LoR*, II, 54; III, 520, 524.
'Working-class': Rosebury, *Tolkien*, 75–6.
Posh tones: the last point is made by Colebatch, *Return*, 64.

Exactly nine modern Ringwraiths: joke.

Hour of the Shire-folk: *LoR*, I, 354.

Accusation: for a recent and typically thoughtless repetition, see Roz Kaveney, 'The Ring recycled,' *New Statesman & Society* (20/27.12.91), 47, who also associates Tolkien with 'a broadside attack on modernism and even on realism' (is *nothing* sacred?), and anachronistically blames him for current 'American commercial fantasy and science fiction.'

31.

Evil creatures: *LoR*, II, 14, 357.

'Moral cartography': Scheps, 'Fairy-tale,' 44–5; also 46, for discussion of the instances of blackness.

Counter-examples: see Rosebury, *Tolkien*, 79.

Kathleen Herbert, *Spellcraft: Old English Heroic Legends* (Hockwold-cum-Wilton, Norfolk: Anglo-Saxon Books, 1993) 271, 225.

Luling, 'Anthropologist.' However, as she adds, the Orcs — as distinct from the Haradrim, Variags and Easterlings — 'are a separate problem, and one that Tolkien himself never really solved' (p. 56); see J.R.R. Tolkien, *Morgoth's Ring*, ed. Christopher Tolkien (London: HarperCollins, 1993), for his efforts to do so.

32.

Stimpson, *Tolkien*, 18.

'Kingly Gondorians': Attebery, *Strategies*, 33.

Brown-skinned: *LoR*, III, 229.

Le Guin, *Language*, 57–8; and see Attebery, *Strategies*, 33, and Rosebury, *Tolkien*, 75–6. (This is simplistic?)

A tale of this sort: *Letters*, 212.

Kilby, *Tolkien*, 51–2.

33.

North-west Europe: *Letters*, 375–6.

Raymond Williams, *The Country and the City* (London: Hogarth Press, 1985), 258. See also Alan O'Connor, *Raymond Williams: Writing, Culture, Politics* (Oxford: Basil Blackwell, 1989). Two disclaimers: I note and appreciate Williams' opening-out of critical vistas from the confines of Leavisism. And I do not mean to subsume Marxism in the work of Williams; there are others, especially Adorno and Horkheimer of the Frankfurt School, who were deeply sceptical about Enlightenment rationalism.

34.

Pastoral: John Lucas, *England and Englishness* (London: Hogarth Press, 1990), 118; John Barrell and John Bull, the editors of *The Penguin Book of English Pastoral Verse* (1974), 5–8.

Williams, *Country*, 36–7, 247.

'The material': see William H. Sewell, Jr, 'Towards a Post-Materialist Rhetoric for Labour History,' 16–23 in Lenard R. Barlanstein (Ed.), *Rethinking Labour History* (Urbana: University of Illinois Press, 1993).

Political character: see Ernesto Laclau and Chantal Mouffe, *Hegemony and Socialist Strategy. Toward a Radical Democratic Politics* (London: Verso, 1985), and their subsequent work.

35.

Ideas, values etc.: a mistake Gramsci never made (nor, for that matter, Mrs Thatcher). To be fair, this is something that the best of Williams' former students, such as Stuart Hall, absorbed and have themselves said.

Suburb: *Letters*, 65.

Shire: *LoR*, I, 120; III, 174.

No havens: Patrick Grant, 'Tolkien: Archetype and Word,' 87–105 in Neil D. Isaacs and Rose A. Zimbardo (Eds.), *Tolkien: New Critical Perspectives* (Lexington: University Press of Kentucky, 1981), 99.

World-view: Williams' *Towards 2000* (London: Chatto & Windus, 1983) firmly subordinates ecology to this logic.

E. P. Thompson, *William Morris*, second edn. (New York: Pantheon, 1976); *Witness Against the Beast: William Blake and the Moral Law* (Cambridge: Cambridge University Press, 1993).

36.

Thompson's garden: from the *New Left Review* 102 (Sept/Oct 1993).

Orwell: quoted by David Ehrenfeld in his excellent essay 'The Roots of Prophecy: Orwell and Nature,' 8–28 in *Beginning Again: People and Nature in the New Millennium* (Oxford: Oxford University Press, 1993), 25. In a letter of 20 July 1933 (quoted by Ehrenfeld on p. 21), Orwell wrote — and note the order of personal priorities — 'The heat here is fearful but it is good for my marrows and pumpkins, which are swelling almost visibly. We have had lashings of peas, beans just beginning, potatos rather poor, owing to the drought. I have finished my novel [*Burmese Days*].'

Williams, *Country*, 35–6.

37.

Fred Inglis, *The Promise of Happiness: Value and Meaning in Children's Fiction* (Cambridge: Cambridge University Press, 1981), 197.

Fred Inglis, 'Gentility and Powerlessness: Tolkien and the New Class,' 25–41 in Robert Giddings (ed.), *J.R.R. Tolkien: This Far Land* (London: Vision Press, 1983), 40. And this despite his 'grateful' acceptance of Claude Rawson's correction on this point in the *Times Literary Supplement* of 26.7.82 — 'Gentility,' 41, n. 18. Inglis's essay on Tolkien is one of the most disingenuous and malicious in a crowded field.

Victory: Rev. Ian A. Muirhead, 'Theology in Gandalf's Garden,' *Arda* (1986), 14–24: 20.

Wagner: *Letters*, 306; see Shippey, *Road*, 296.

Ragnarok: Brian Branston, *The Lost Gods of England* (London: Thames and Hudson, 1957), 155.

'Patriotism': *Letters*, 63–4.


38.

Not a socialist: *Letters*, 235.

Fascism: for a fuller discussion of Tolkien's opposition to fascism, see Robert Plank, '"The Scouring of the Shire": Tolkien's View of Fascism,' 107–15 in Lobdell, *Compass;* and Jessica Yates, 'Tolkien the Anti-Totalitarian,' 233–45 in Reynolds and GoodKnight, *Proceedings.*

Modernists: see John R. Harrison, *The Reactionaries* (London: Victor Gollancz, 1966).

Technological modernism: see Zygmunt Bauman, *Modernity and the Holocaust* (Cambridge: Polity Press, 1989); Stanley G. Payne, *A History of Fascism, 1914–1945* (Madison: University of Wisconsin Press, 1996); and Jeffrey Herf, *Reactionary Modernism* (Cambridge: Cambridge University Press, 1984).


39.

Grudge: *Letters*, 55.

German translation: *Letters*, 37.

Half republic: *Letters*, 241.


40.

City: *LoR*, III, 296–7. See Jason Finch, 'Democratic Government in Middle-Earth,' *Amon Hen* 129 (1994), 12–13; and Madawc Williams, 'Good Government in Middle-Earth,' *Amon Hen* 132 (1995), 17, 19.

Republicanism: see Patrick Curry, *Machiavelli for Beginners* (Cambridge: Icon Books, 1995), for an introduction and bibliography.

Donald Davie, *Thomas Hardy and British Poetry* (London: Routledge & Kegan Paul, 1973), 93–4.

Canadians: see Mathis Wackernagel, 'How big is our ecological footprint?,' *Real World* No. 16 (Summer 1996), 7–9.


41.

Mordor: *LoR*, III, 240; *Letters*, 154.

Madawc Williams, 'Good Government,' 17.

Resistance to fascism: see Plank, 'The Scouring of the Shire'; he also rightly points out that 'Tolkien opposes fascism as a conservative rather than as a democrat' (p. 114).


42.

Bate, *Romantic Ecology*, 46.

Bate, *Romantic Ecology*, 11.

One small garden: *LoR*, III, 210–11.

**43.**

Harrison, 'England,' 170–71. *Cf.* Robert Pogue Harrison, *Forests: The Shadow of Civilization* (Chicago: University of Chicago Press, 1992), 156: 'nostalgia keeps open the vision of historical alternatives . . .'

Dartmoor: quoted by Meredith Veldman, *Fantasy, the Bomb, and the Greening of Britain: Romantic Protest, 1945–1980* (Cambridge: Cambridge University Press, 1994), 110. On recent direct-action movements in the UK, see George McKay, *Senseless Acts of Beauty* (London: Verso, 1996).

**44.**

Veldman, *Fantasy*, shows that the connections between Tolkien's popularity and the CND/END protest movement far outweigh E. P. Thompson's superficial use of imagery from Tolkien's books to describe a Cold War mentality — something about which he was corrected (as he later acknowledged) by Jessica Yates; see her 'Tolkien the Anti-Totalitarian.'

Taggart quoted in Veldman, *Fantasy*, 108.

**45.**

Maria Kamenkovich, 'The Secret War and the End of the First Age: Tolkien in the (former) USSR,' *Mallorn* 29 (1992), 33–38: 36, 38. See also Vladimir Grushetskiy, 'How Russians See Tolkien,' 221–25, and Natalia Grigorieva, 'Problems of Translating into Russian,' 200–205 in Reynolds and Good-Knight, *Proceedings*; and 'Tolkien fantasies strike Russian Chord,' *The Globe and Mail* (28.5.94).

**46.**

Escapism: 'Fairy-Stories,' 56.
Motor cars: 'Fairy-Stories,' 57.
Progressive things: 'Fairy-Stories,' 58.

**47.**

*LoR*, III, 378; my thanks to Nicola Bown for this interpretation of Sam's words.

Geoffrey Grigson, *Collected Poems 1963–1980* (Alison & Busby, 1982), 198.

## 3. Middle-earth: Nature and Ecology

**48.**

Epigraph: Joni Mitchell, from 'Dog Eat Dog' (1985).
'Beyond community': Kane, *Wisdom*, 190.
Time and place: *Letters*, 283, 239.

49.
Middle-earth: *Letters*, 283.
'Tolkien's readers': Dwayne Thorpe, 'Tolkien's Elvish Craft,' 315–21 in *Proceedings*, 315. See too Janet Menzies, 'Middle-Earth and the Adolescent,' 56–72 in Giddings, *This Far Land*, 57; Le Guin, *Language*, 148; Lee D. Rossi, *The Politics of Fantasy: C. S. Lewis and J.R.R. Tolkien* (Ann Arbor: UMI Research Press, 1984), 128.
Barbara Strachey, *Journeys of Frodo: An Atlas of J.R.R. Tolkien's The Lord of the Rings* (London: George Allen & Unwin, 1981), who notes in her Foreword that 'The evidence is — as one might expect — splendidly consistent.' And not only internally; for example, we learn from Map 27 that Aragorn, Gimli and Legolas covered 135 miles in their four-day chase of the orcs, or 33¾ miles a day. That is closely equivalent to the advance of Alexander the Great's army at its swiftest: an average of 36 miles a day, in one campaign, or between 18 and 20 miles a day under full armament. (Information on Alexander from *The New Yorker*, 9.12.91, p.130.)

50.
Profound presence: *cf.* Ezard, 'Tolkien's Shire': 'My overriding memory of Tolkien is not of his hobbits or his elf-queens (these can pall as a reader grows up) but of his sense of stewardship for the unperverted natural world.'

51.
Elves: *LoR*, I, 468.
Wanderers: a point made by William Dowie in 'The Gospel of Middle-Earth according to J.R.R. Tolkien,' 265–85 in Salu and Farrell, *Tolkien*, 270.
Paul Kocher, 'Middle-Earth: An Imaginary World?,' 117–32 in Isaacs and Zimbardo, *Tolkien*, 125.
Caras Galadhon: *LoR*, I, 457, 458.

52.
Colours: *LoR*, II, 86–7, 413; III, 276, 163; II, 39.
Angela Carter, *Black Venus* (London: Chatto and Windus, 1985), 67–9.

53.
Shakespeare: *LoR*, III, 529.
Dark hearts of trees: *LoR*, I, 179.
Eärendil: I am referring here to Tolkien's first and best formulation, before his destructive self-doubts of the late 1950s; see *Morgoth's Ring*, 197–99 and *passim*.

54.
Whose side: *LoR*, II, 89.
Tree-love: *Letters*, 257, 220.

Totem tree: Kilby, *Tolkien*, 21. *Letters*, 419–20.

Figures on loss of UK woodlands: Council for the Protection of Rural England (CPRE), reported in *The Observer* 25.8.85.


55.

Figures for European woodlands: *The European* (6–9.5.93). For tropical forests: Robert Lamb, *World Without Trees* (London: Methuen, 1979); Nigel Dudley, *The Death of the Trees* (London: Pluto Press, 1985).

Foreword by John Fowles (xiii–xv) in Angela King and Sue Clifford, *Trees Be Company: An Anthology of Poetry* (Bristol: Bristol Classical Publishing, 1989), xiv.


56.

Bate, *Romantic Ecology*, 40.

Seneca: quoted in Anne Bancroft, *Origins of the Sacred* (London: Arkana, 1987), 101–2. See also Yvonne Aburrow, *The Enchanted Forest: The Magical Lore of Trees* (Chieveley, Berks.: Capall Bann, 1993); Anand Chetan and Diana Brueton, *The Sacred Yew* (London: Penguin/Arkana, 1994); J. H. Philpot, *The Sacred Tree* (Felinfach: Llanerch, 1994 [1897]).


57.

R. L. Stevenson, *Forest Notes* (1876), quoted in Kim Taplin, *Tongues in Trees: Studies in Literature and Ecology* (Bideford: Green Books, 1989), 15.

Jay Griffiths, 'The dying fall,' *The Guardian* (14.2.96).

Universal call: in addition to the work of Eliade, see Marina Warner, 'Signs of the Fifth Element,' 7–47 in *The Tree of Life: New Images of an Ancient Symbol* (London: The South Bank Board, 1989).

Urban trees: see Richard Mabey, *Flora Britannica* (London: Sinclair-Stevenson, 1996).

Mircea Eliade, *Patterns in Comparative Religion* (London: Sheed and Ward, 1958), 267.


58.

The Apple-tree: H. R. Ellis Davidson, *Myths and Symbols in Pagan Europe: Early Scandinavian and Celtic Religions* (Manchester: Manchester University Press, 1988), 170; *LoR*, I, 243.

Eliade, *Patterns*, 269, 149.

Mythical tree: *Letters*, 321; 342.


59.

John L. Peyton, *The Birch: Bright Tree of Life and Legend* (Blacksburg VA: McDonald & Woodward, 1994).

John Fowles's superb essay *The Tree* (St Albans: The Sumach Press, 1992; first published London: Aurum Press, 1979), 80.

Tolkien, *Tree and Leaf*, 6.

Taplin, *Tongues*, 196.
Old Forest: *LoR*, I, 179.

60.
Saruman: *LoR*, II, 90–91, 89.
The Ring: *LoR*, III, 184–5; I, 317.

61.
Allegory/orcs: *LoR*, I, 11–12; *Letters*, 262.
The One Ring: *Letters*, 177.
Goblins: *Hobbit*, 69.

62.
Vulgar devices: 'Fairy-Stories,' 15.
Magic *vs.* enchantment: 'Fairy-Stories,' 18, 49; *Letters*, 151–2; 'Fairy-Stories,' 49–50; *Letters*, 146, 200.

63.
Evil: *Letters*, 146, 151; *LoR*, I, 339.
Magic into science: e.g., Charles Webster, *From Paracelsus to Newton: Magic and the Making of Modern Science* (Cambridge: Cambridge University Press, 1982); and for an account of a small part of this process in seventeenth- and eighteenth-century England, see my *Prophecy and Power: Astrology in Early Modern England* (Cambridge: Polity Press, 1989).
Giant Bomber: *Letters*, 88.

64.
Elves/Noldor: *Letters*, 236, 190.
Mills/orc-work/dwarves: *Letters*, 200; *LoR*, II, 107; *LoR*, III, 529; II, 189.

65.
Gollum: Thanks to Jesper Siberg for this point.
Power of Ring: *LoR*, I, 347.
Bombadil: U. Milo Kaufmann, 'Aspects of the Paradisiacal in Tolkien's Work,' 143–52 in Lobdell, *Compass*, 151; and Charles E. Noad, 'The Natures of Tom Bombadil: A Summary,' 79–83 in T. A. Shippey *et al.*, *Leaves from the Tree: J.R.R. Tolkien's Shorter Fiction* (London: The Tolkien Society, 1991), 83. (Personally, I find Bombadil's verse and talk very trying, and Goldberry wholly unbelievable.) *Letters*, 26. *LoR*, I, 348.

66.
Mordor: *LoR*, II, 296–7.
Morgul Valley/Mount Doom: *LoR*, II, 402; III, 261.
Shippey, *Road*, 124, 126.

67.
Malevolent amalgam: see Ekins, *New World Order*, from whom I have borrowed this analysis.
Machines: *Letters*, 11.
Long defeat/Barad-dûr: *LoR*, I, 463; II, 199. (I have changed the tense from past to present.)
Addictive indeed: see Helena Norberg-Hodge, *Ancient Futures: Learning from Ladakh* (London: Century, 1992). (This 'twist' is properly called hegemony.) See also the work of Andre Gorz and Michael Walzer, on freeing spheres of life from the inappropriate and destructive domination of economic logic. (Just as a tiny exercise, imagine, for a moment, being in a public space — which may now itself be a difficult feat, in some places — without being screamed at, bullied, enticed, intimidated, patronized and lied to by advertising. What bliss!)

68.
Shippey, *Road*, 78.
Michel Foucault, 'Nietzsche, Genealogy, History,' 76–100 in Paul Rabinow (Ed.), *The Foucault Reader* (Harmondsworth: Penguin, 1984), 96. Jane Chance too, in *The Lord of the Rings: The Mythology of Power* (New York: Twayne Publishers, 1992), has noticed the affinity between the Eye of Sauron and (via Bentham's Panopticon) Foucault's model of knowledge/power.
Teller: Christopher Hitchens, 'Dr Strangelove, I presume?' *New Statesman & Society* (30.9.94), 44–5: 45.

69.
Max Horkheimer and Theodor W. Adorno, *The Dialectic of Enlightenment* (New York: Continuum, 1994 [1944]), 28.
On Descartes: Philip J. Davis and Reuben Hersh, *Descartes' Dream: The World According to Mathematics* (Harmondsworth: Penguin, 1986) 3–4.
Horkheimer and Adorno, *Dialectic*, 4, 7, 5, 18.
Mordor as industrial: e.g., Nick Otty, 'The Structuralist's Guide to Middle-Earth,' 154–78 in Giddings, *This Far Land*, 166.

70.
Primo Levi, *If This is a Man/The Truce* (London: Sphere Books, 1987), 48, 78. See also Bauman, *Modernity*.

71.
Regeneration: see, for example, 'Reaping their own rewards,' *The Guardian* (13.11.1996).

72.
Fowles, *Tree*, 82.

Hedgerows through hay-meadows: statistics from *Wildfowl & Wetlands* 115 (Spring 1996); the C.P.R.E.; and Plantlife.
Always a-hammering: *LoR*, III, 356.


73.
Endangered species: *The Guardian* (25.1.95 and 2.8.95).
Cars: *The Observer* (8.10.94).
Water: John Vidal, 'The water bomb,' *The Guardian* (8.8.95).
Sea: Sylvia A. Earle, *Sea Change* (London: Constable, 1996).
Human population: *The State of World Population* (New York: U.N. Population Fund, 1995).
Biological catastrophe : Richard Leakey and Roger Lewin, *The Sixth Extinction: Biodiversity and its Survival* (London: Weidenfeld and Nicolson, 1996); and see Colin Tudge, *The Day Before Yesterday: Five Million Years of Human History* (London: Jonathan Cape, 1996).
Peter Raven of the Missouri Botanical Gardens, quoted in *The Guardian* (24.2.94).


74.
David Ehrenfeld, *The Arrogance of Humanism* (Oxford: Oxford University Press, 1978) — a powerful book. On technological *hubris*, Ehrenfeld was anticipated by Bateson, *Steps*, 426–40; and followed by Edward Tenner, *Why Things Bite Back: Technology and the Revenge of Unintended Consequences* (New York: Alfred A. Knopf, 1996).
Michel de Montaigne, *The Complete Essays* (London: Penguin Books, 1993), 1133. There are contemporary voices of wisdom and sanity, too, of course: Bateson, Feyerabend, Bauman and others mentioned or quoted herein.
Utter folly: *Letters*, 116.


75.
DNA: David Suzuki, 'Lifebusters,' *BBC Wildlife* (Sept. 1993), and John Vidal and John Carvel, 'Lambs to the gene market,' *The Guardian* (12–13.11.94).
Life plc (Public Limited Company): for a good recent discussion see John Frow, 'Information as Gift and Commodity,' *New Left Review*, 219 (1996), 89–108. For examples of recent apologias for this brave new world, see those of Gregg Easterbrook and Richard D. North. (The Ring needs, and finds, servants to wield words as much as guns.)
Many defeats: *LoR*, I, 318.
George Orwell, in the *Partisan Review* (July–August 1947).


76.
Mushrooms: *LoR*, I, 143.
'Wild fungi face extinction as pollution threat increases' — *The Observer* (29.12.91); and *BBC Wildlife* (Feb. 1995).

Richard Mabey, 'Introduction: Entitled to a View?,' in Mabey, *Second Nature*, ix–x.

77.
Harrison, 'England,' 168–9. For an unflinching portrayal of the contemporary cultural and natural asset-stripping of 'the heart of England,' see V. S. Naipaul, *The Enigma of Arrival* (London: Viking, 1985).
Common Ground: King and Clifford, *Trees Be Company*, ix. See also Mabey *et al.*, *Second Nature*; Richard Mabey, in Neil Sinden, *In a Nutshell: A Manifesto for Trees and a Guide to Growing and Protecting Them* (London: Common Ground, 1989), which is 'a guide to how we might join imaginative responsibility with practical understanding to reach a *modus vivendi* with trees' (p. 5); and see Herbert Whone, *Touch Wood: A Journey Among Trees* (Otley: Smith Settle, 1990). For a fair and intelligent discussion of Common Ground, see Patrick Wright, 'Lexicon of life for the common man,' *The Guardian* (4.7.92).
John Ruskin, *Unto This Last*, with an Introduction by J. A. Hobson (London: Cassell & Co., 1907 [1862]), 154.

78.
Keith Thomas, *Man and the Natural World: Changing Attitudes in England 1500–1800* (London: Allen Lane, 1983), 251, 267, 287.
Ludmilla Jordanova, 'The Interpretation of Nature: A Review Article,' 195–200 in *Comparative Studies in Society and Nature* 29:1 (January 1987) [rev. of Thomas, *Man*], 198, 200.

79.
Trees: let me risk feminist wrath and add that only women and music can offer any serious competition (which, for all I know, may be because of something they all share).
Overdevelopment: see Henry Porter, 'Heart acres,' *Guardian* (16.9.96). Noise: see research published by the C.P.R.E. in 1996.
Sanity and sanctity: Carole Batten-Phelps, quoted in *Letters*, 413.

80.
Wise Use: see Andy Rowell, *Green Backlash* (London: Routledge, 1996).

81.
R. P. Harrison, *Forests*, 121, 123, 124.
Fowles, *Tree*, 43–4.

82.
Trees: *LoR*, I, 172; I, 456.

178 ❦  *References*

Baja: Luling, 'Anthropologist,' 54, quoting Louis Sarno, *Song from the Forest* (London: Bantam Press, 1993).

R. Limaugh: quoted in *Guardian* (7.4.95).

United States: furthermore, it seems, 'God put the owl here for man.' Excuse me? From 'Loggers vote; spotted owls and trees can't,' *BBC Wildlife* (July 1992), 49–50.

Industrial forestry: Herb Hammond, *Seeing the Forest among the Trees* (Vancouver: Pole Star Press, 1991) — quoted in *The Ecologist* 24:1 (1994), 36–7.


83.

Hog Farmer: quoted in Alexander Cockburn, 'A Short, Meat-Oriented History of the World: From Eden to Mattole,' *New Left Review* 215 (1996), 16–42: 20.

Indigenous peoples: Fred Pearce, 'Ripe plywood,' *BBC Wildlife* (January 1991). *LoR*, II, 187, 189.


84.

New communion: Fraser Harrison, *The Living Landscape* (London: Mandarin, 1991), 13.

G. Robinson, in Hammond, *Seeing the Forest* (see note on p. 82 above).

Field/'chemicals' etc.: 'Just a field,' *Guardian* (12.10.94). See *Department of the Environment Countryside 1990 Series 2* (1993); Marion Shoard, *This Land is Our Land* (London: Paladin, 1987); and the work of the Open Spaces Society and the Countryside Restoration Trust.


85.

'Green and living': Bill Bryson, *Notes from a Small Island* (Toronto: Reed Books, 1995), 132.

Mill: Ezard, 'Tolkien's Shire.'

Richard Mabey, *In a Green Shade* (London: Unwin, 1985), 12. *Cf. The Tree of Life: New Images of an Ancient Symbol* (London: The South Bank Board, 1989), Warner, 'Signs,' 43: '. . . on the whole, the pastoral mode arises out of acquired knowledge, not inborn experience, and is the work of people who are strangers in nature, trying to learn.'

Urban spaces: Jacquelin Burgess, Carolyn M. Harrison and Melanie Limb, 'People, Parks and the Urban Green: a study of popular meanings and values for open spaces in the city,' *Urban Studies* 25 (1988), 1–19 (courtesy Colin Ward). *Cf.* Ken Walpole, *Park Life: Urban Parks and Social Renewal* (London: Demos, 1995).

William Hazlitt, 'On the Love of the Country,' 3–8 in *Selected Essays of William Hazlitt*, ed. Geoffrey Keynes (London: Nonesuch Press, 1970), 7.

Scientific case: see Neil Evernden, *The Natural Alien* (Toronto: University of Toronto Press, 1985), and R. P. Harrison, *Forests;* although they were anticipated in this respect by Ehrenfeld, *Arrogance*.

86.

F. Harrison, 'England,' 172.

F. Harrison, *Landscape*, 13.

## 4. The Sea: Spirituality and Ethics

87.

Epigraph: Alkis Kontos, 'The World Disenchanted, and the Return of Gods and Demons,' 223–47 in Asher Horowitz and Terry Maley (Eds.), *The Barbarism of Reason: Max Weber and the Twilight of Enlightenment* (Toronto: University of Toronto Press, 1994), 228.

The Sea: *LoR*, III, 178, 321 (and 378).

88.

Stars: *LoR*, III, 238; and see *LoR*, II, 270.

Kontos, 'The World Disenchanted,' 225.

Escapism: 'Fairy-Stories,' 60–61.

Giddings, *This Far Land*, 12–13. *Cf.* the similar point made by Rosemary Jackson, *Fantasy: The Literature of Subversion* (London: Methuen, 1981), 154–5.

89.

Uprooting the evil: *LoR*, III, 185.

Levi, *If This is a Man*, 188.

William Empson, *Some Versions of Pastoral* (London: Chatto & Windus, 1979), 4–5.

Le Guin, *Language*, 100.

90.

Warner: *Independent* (3.2.94); *cf.* 'Children are our copy, in little . . .' (*Independent*, 10.2.94); then quoted by Henry Porter, 'Reason eclipsed by evil,' *Guardian* (17.3.96) (my italics).

Le Guin, *Language*, 58–9.

D. J. Enright, *Interplay: A Kind of Commonplace Book* (Oxford: Oxford University Press, 1995), 160.

91.

My story: *Letters*, 246, 262.

Immortality/endless living: *Letters*, 267; 'Fairy-Stories,' 62.

92.

Reject death: *LoR*, III, 425; Fowles, *Tree*, 81. *Cf.* Robert D. Romanyshyn, *Technology as Symptom and Dream* (London: Routledge, 1984).

Timothy Leary, *Neuropolitics* (1977).

Natural span: *Letters*, 155. See also Malcolm Gladwell, 'The New Age of Man,' *The New Yorker* (30.9.1996).

93.
C. Winter: quoted in *Daily Mail* (18.7.96); and see Jonathan Romney's report in *Guardian* (31.10.96).
J.R.R. Tolkien, 'Beowulf: The Monsters and the Critics,' in Donald K. Fry (Ed.), 8–56 in *The Beowulf Poet: A Collection of Critical Essays* (Englewood Cliffs: Prentice-Hall, 1968) (originally *Proceedings of the British Academy* XXII [1936] 245–95), 25, 31.
Seed of courage: *LoR*, I, 192.


94.
Tridentine: Kilby, *Tolkien*, 53.
Chance-meeting: *LoR*, I, 84; III, 450.
Shippey, *Road*, 138, 148.


95.
Ellis Davidson, *Myths*, 221–23.
Alby Stone, *Wyrd: Fate and Destiny in North European Paganism* (Wymeswold: Heart of Albion Press, 1989), 23, 21.
Chilling: *Letters*, 330–33.
Crack of Doom/The Quest: *LoR*, III, 268; I, 89; III, 271.


96.
A Christian work: for a good explication of Christian imagery in *LoR*, see Gracia Fay Ellwood, *Good News from Tolkien's Middle Earth* (Grand Rapids: Wm. B. Eerdmans Publ. Co., 1970), 87–142. On pre-Christian religion, see Stone's excellent *Wyrd*.
Happy Ending: 'Fairy-Stories,' 62.
Kath Filmer, *Scepticism and Hope in Twentieth Century Fantasy Literature* (Bowling Green: Bowling Green State University Popular Press, 1992), 27.


97.
No true end: 'Fairy-Stories,' 62.
*LoR*, III, 378; *Letters*, 328. (Italics in the original.) *Cf.* Filmer, *Scepticism*, 29–30. 'Fairy-Stories,' 62.
Natural theology/religious and Catholic: *Letters*, 220, 235, 172.


98.
The One: *Letters*, 284, 235.
Reshaping: *LoR*, III, 385.
The Valar: *LoR*, III, 382, 489.
Active animism: *Hobbit*, 205, 167; *LoR*, I, 383; III, 168, 432.


99.
Minas Tirith, etc.: *LoR*, III, 121 (and see Shippey, *Road*, 193–4); *LoR*, III, 142; *LoR*, III, 192.

Ruskin quoted in Bate, *Romantic Ecology*, 61, 83.
Midsummer Eve: *LoR*, III, 303.


100.

Wonderful sunshine: *LoR*, III, 369.
Primary World: *Letters*, 187, 189. See also Tolkien, *Morgoth's Ring*, 361–6, for his further thoughts on Elvish reincarnation.
T. A. Shippey, 'Long Evolution: *The History of Middle-Earth* and its Merits,' *Arda* (1987), 18–39: 32.


101.

Gandalf/Wodan: Branston, *Lost Gods*, 86, 91–2; Roger Lancelyn Green, *Myths of the Norsemen* (Harmondsworth: Penguin, 1970), 203; H. A. Guerber, *Myths of the Norsemen* (London: George G. Harrap & Co., 1908), 16–17.
H. R. Ellis Davidson, 'Scandinavian Cosmology,' 172–97 in Carmen Blacker and Michael Loewe (Eds.), *Ancient Cosmologies* (London: George Allen & Unwin, 1975), 186.
'Odinic wanderer': *Letters*, 119.


102.

Kilby, *Tolkien*, 57–8; and Shippey, *Road*, 218–19. (But note Tolkien was not necessarily consistent on the question of his mythology's origins.)
Radagast: I owe this wonderful point to Skrét Nekulová.


103.

Zipes, *Magic Spell*, 146.
Hobbits: *Letters*, 158. *Cf.* another statement that 'Elves are certain aspects of Men . . .' (p. 189).
Grant, 'Archetype,' 101–2.
See Shippey, *Road*, 179. On 'virtuous pagans' see John Casey, *Pagan Virtue: An Essay in Ethics* (Oxford: Clarendon Press, 1990). Gunnar Urang, in *Shadows of Heaven: Religion and Fantasy in the Writing of C. S. Lewis, Charles Williams, and J.R.R. Tolkien* (Philadelphia: Pilgrim Press, 1971), describes Tolkien as engaged in 're-paganizing' (p. 120).
Kane, *Wisdom*, 238.


104.

Russell Hoban, *The Moment Under the Moment: Stories, a Libretto, Essays and Sketches* (London: Jonathan Cape, 1992), 138; 146. There are several significant overlaps of substance if not of style between Tolkien's and Hoban's work, although I doubt that the latter, influenced by psychoanalysis and classical Greek mythology, would admit it. (Isn't it strange how there always seems to be some reason — usually not a very good one — for disavowing Tolkien? And how the determination to be Adult has such childish consequences?)

The law: 'Fairy-Stories,' 99.

105.
P. L. Travers, *What the Bee Knows: Reflections on Myth, Symbol and Story* (London: Penguin/Arkana, 1989), 297.
Soup: a metaphor I have borrowed from Tolkien's 'On Fairy-Stories.'

106.
'Beowulf,' 25–6; *cf.* Shippey, *Road*, 179–80.

107.
Every major world religion: see J. Baird Callicott, *Earth's Insights* (LA/Berkeley: University of California Press, 1990, and Roger S. Gottlieb (Ed.), *This Sacred Earth* (New York: Routledge, 1996), for surveys.

108.
Celtic Christianity: see, e.g., Esther de Waal, *The Celtic Way of Prayer: The Recovery of the Religious Imagination* (London: Hodder and Stoughton, 1996).
Genesis/Acts, etc.: see Lynn White, 'The Roots of our Ecologic Crisis,' *Science* 155 (1967), 1203–7, reprinted in Ian Barbour (Ed.), *Western Man and Environmental Ethics* (Reading MA: Addison-Wesley, 1973). See also Eugene C. Hargrove (Ed.), *Religion and Environmental Crisis* (Athens GA: The University of Georgia Press, 1986), especially his Introduction, 'Religion and Environmental Ethics: Beyond the Lynn White Debate,' ix–xix.
Kane, *Wisdom*, 255.
Weber: see Kontos, 'The World Disenchanted.'

109.
'Pagans' (or alternatively 'Heathens') includes practitioners of Wicca and witchcraft, Odinism, Shamanism, Druidism, Eco-paganism and Eco-feminism.
'Collective spirituality': this term was used by a British anti-roads protester in an article I read but cannot now locate; I wish I could, to thank him, because it is perfect.
'Elective affinity': Graham Harvey, 'The Roots of Pagan Ecology,' *Religion Today* 9:3 (1994), 38–41: 40–41; Mary Dann, Western Shoshone Nation, in *Cornerstones* (Oct. 1994).
Humanism: see Ehrenfeld, *Arrogance.*

110.
C. S. Lewis: quoted in Shippey, *Road*, 68–80.
Richard L. Purtill, *J.R.R. Tolkien: Myth, Morality, and Religion* (San Francisco: Harper & Row, 1984), 3, 5–6. *Cf.* Jared Lobdell, *England and Always: Tolkien's World of the Rings* (Grand Rapids: William B. Eerdmans Publ. Co., 1981), 81; and Shippey, *Road*, 148. In an interview in 1966, Tolkien said of ancient

myths: 'I tried to improve on them and modernize them. To modernize them is to make them credible.' From Henry Resnik, 'An Interview with Tolkien,' *Niekas* 18 (Spring 1967), 37–47: 40. (With thanks to Charles Noad.)

Zipes, *Spell*, 146.

Randel Helms, *Tolkien's World* (London: Granada/Panther, 1976), 23.

Arthurian myth: *Letters*, 144 (my italics).

111.

Heroic world: *LoR*, III, 174; a point made by Lobdell, *England*, 85. 'Myth/truth': Shippey, *Road*, 188.

Pagan/Christian: see too Christopher Milne, *The Hollow on the Hill: The Search for a Personal Philosophy* (London: Methuen, 1982).

The West: *Letters*, 186; 156.

Straight sight: *LoR*, III, 393.

## 5. *Fantasy, Literature and the Mythopoeic Imagination*

112.

Epigraph: Aristotle, *Metaphysics*, A, 2, 982 b18.

'Evidently missing': Shippey, 'Evolution,' 23–4. See also Anders Stenström, 'A Mythology? For England?,' 310–14 in Reynolds and GoodKnight, *Proceedings*.

113.

'Pull of faërie': Helen Armstrong, in *Amon Hen* 139 (1996), 13–14.

'Fairy-Stories,' 44 (a misunderstanding he traces in, if not to, the *Oxford English Dictionary*) and 13–14 (he adds: 'all things of which children have, as a rule, less need than older people').

114.

Setting off: *LoR*, I, 490.

Lothlórien/Frodo on Cerin Amroth: *LoR*, I, 457, 455.

115.

Brewer, 'Romance,' 261–2. The same point is made by Geoffrey Ridden, *Notes on The Lord of the Rings* (Harlow, Essex: Longman York Press, 1984).

116.

Taplin, *Tongues*, 91. (This is a question I cannot expand upon here, but do in my paper — see Preface.)

Edward Blishen, 'Town: Bad, Country: Good,' 15–24 in Mabey et al., *Second Nature*, 24.

Kilby, *Tolkien*, 79. (A sixth-grade pupil would be about twelve years old.)

117.
Pain and delight: *LoR*, III, 280.
'The ongoing "What if?"': Hoban, *Moment*, 154.
Italo Calvino, *Italian Folk Tales*, transl. George Martin (New York: Pantheon Books, 1980 [1956]), xviii–xix.

118.
'Lies breathed through silver': C. S. Lewis's description, to which Tolkien replied in his poem 'Mythopoeia,' published in *Tree and Leaf.*
Carl Kerenyi, 'Prolegomena,' 1–32 in C. G. Jung and C. Kerenyi, *Introduction to a Science of Mythology: The Myth of the Divine Child and the Mysteries of Eleusis* (London: Routledge & Kegan Paul, 1951), 2, 3–4, 8.

119.
Milton Scarborough, *Myth and Modernity: Postcritical Reflections* (Albany: SUNY, 1994), 110.
Luling, 'Anthropologist,' 56.

120.
Lobdell, *England*, 88. Robert Giddings and Elizabeth Holland, *J.R.R. Tolkien: The Shores of Middle-Earth* (London: Junction Books, 1981), 255, make a similar point: 'To read *The Lord of the Rings* is to walk through the visionary landscapes which generated us' — although for them, the traditions concerned are more Indo-European and Biblical.
Alan Garner, 'A Bit More Practice,' 196–200 in Margaret Meek, Aidan Warlow, Griselda Barton (Eds.), *The Cool Web: The Pattern of Children's Reading* (London: The Bodley Head, 1977), 199. It is all the sadder that Garner evidently feels unable to acknowledge his common concerns with, and clear debts to, Tolkien; or even our debt generally, as readers.

121.
Discovery: *LoR*, I, 9.
Tolkien, 'Beowulf,' 19.

122.
Calasso, *Marriage*, 281.

123.
Kane, *Wisdom*, 45, 34.
Tolkien's Elves: thanks to Nicola Bown for this point.

124.
Christopher Tolkien: *Morgoth's Ring*, 371–72.
Glorfindel: *LoR*, I, 279, 286.
High Elves: *LoR*, I, 292.

125.
Philip Wheelwright, 'Poetry, Myth, and Reality,' in Gerald J. Goldberg and Nancy M. Goldberg (Eds.), *The Modern Critical Spectrum* (Englewood Cliffs: Prentice-Hall, 1962): 319.


126.
Hoban, *Moment*, 139.
Calasso, *Marriage*, 94.
Travers, *What the Bee Knows*, 209. (I think she is quoting James Hillman.)
Psychological exploration: see, for example, the excellent work of Ginette de Paris and Jean Shinoda Bolen.


127.
Kane, *Wisdom*, 241.
Definitions: see Shippey, *Road*, 198; Brewer, 'Romance'; and Elgin, *Comedy*, 37.
Story-telling: see Bill Buford, 'The Seductions of Storytelling,' *The New Yorker* (24.6–1.7.96), 11–12.
Nuala O'Faolain, in *Irish Times* (7.11.92).
Amanda Craig, 'Lord of all he conveyed, despite his fans,' *Independent* (25.1.92).


128.
Walter Benjamin, *Illuminations: Essays and Reflections*, ed. Hannah Arendt (New York: Schocken Books, 1969), 'The Storyteller,' 83–109: 83, 87, 89, 91, 101, 102. This essay was first written in 1936. Benjamin's modernist admirers, at least those who have also engaged in Tolkien criticism, seem to have missed it. I am very grateful to Nicola Bown for bringing it to my attention.
'Necessary degree of irony', etc.: Dennis Potter, *Waiting for the Boat: On Television* (London: Faber & Faber, 1984), 12–13, 31. (The depressing degeneration of Italo Calvino's fiction is a characteristic case in point.)
Starving audience: *Letters*, 209. This is the need for story, unacknowledged but urgent, that Philip Pullman recently described; see my discussion in the Introduction, earlier.
Narrative fantasy: 'Fairy-Stories,' 51.


129.
Attebery, *Strategies*, 40–41, 46.
Fantasy: see Brian Attebery's excellent *Strategies*.
Enright, *Interplay*, 131.
Feeding Gollum: *LoR*, II, 284.
On a 'snide and aggressive' media, see Adam Gopnik, *The New Yorker* (12.12.94), 84–102; also David Nicholson-Lord, 'Write me a novel I can actually read,' *Independent on Sunday* (30.4.95), and Mary Midgeley, 'Sneer tactics,' *Guardian* (7.9.97).

Weber: quoted in Kontos, 'The World Disenchanted,' 233.

130.
Čapek: quoted in *PEN International* 45:1 (1995), 39.
Fowles, *Tree*, 64–5.
W. H. Auden, 'The Quest Hero,' 40–61 in Neil D. Isaacs and Rose A. Zimbardo (Eds.), *Tolkien and the Critics* (Notre Dame: University of Notre Dame Press, 1968), 42.
Proust as 'quest novel': Barry Unsworth, in *PEN International* 44:2 (1994), 43.

131.
Heinrich Zimmer, *The King and the Corpse: Tales of the Soul's Conquest of Evil* (Princeton: Princeton University Press, 1948), 1–3. For the perfect example of a coroner's report, see Frederic Jameson, 'Magical Narratives: Romance as Genre,' *New Literary History*, 7:1 (1975), 135–63.
Ruskin, *Unto This Last*, 145.

132.
*LoR* really is: as was noticed by Colin Wilson, *Tree by Tolkien* (Santa Barbara: Capra Press, 1974), 37: *The Lord of the Rings* 'is at once an attack on the modern world and a credo, a manifesto.'
Bate, *Romantic Ecology*; Elgin, *Comedy*; Veldman, Fantasy. See also the neglected H. W. Piper, *The Active Universe* (London: The Athlone Press, 1962).
Critics' arrogance: another good example is the attempt by modernists like Waldemar Januszczak and Martin Pawley to trash the doubts, fears and dislike of much modern architecture by Prince Charles in 1989–90. They invoked everything from the size of his ears, and supposed associations with Hitler, to economic 'realism' and 'progress.' But the vast outpouring in the media from members of the public in Charles' defence demonstrated that he was, overwhelmingly, speaking for them; and to those feelings, the experts had no convincing reply.

133.
Lindsay was first published in London by Gollancz in 1920. Peake's trilogy — *Titus Groan, Gormenghast*, and *Titus Alone* — was published in 1946–59. (Properly speaking, of course, a 'personal mythology' — like a 'new tradition' — is an oxymoron.)
For a refreshing contrast to Clarke's book, see Randolph Stow, *The Girl Green as Elderflower* (London: Minerva, 1991; first published by Secker and Warburg, 1980).
*Mythago Wood* published in London: Victor Gollancz, 1984; *Little, Big* in New York: Bantam Books, 1981; *The Earthsea Quartet*, published in separate volumes 1968–72 and 1990, is now available as a single volume from Penguin (1993).

A. Dorfman, *The Empire's Old Clothes* (New York: Pantheon Books, 1983), ix.

134.
Former employee: Richard Schickel, *The Disney Version* (New York: Simon & Schuster, 1968), 227. See also Jack Zipes, *Fairy Tale as Myth, Myth as Fairy Tale* (Lexington KY: The University Press of Kentucky, 1994), Chapter 3, 'Breaking the Disney Spell' (pp. 72–95).
Iconic enough: Oliver Bennett, 'Dark side of the toon,' *Observer* (10.9.95).
Cultural capital: see Alexander Cockburn, 'Fatal Attraction,' *The Guardian* (12.5.95).

135.
Doppelgänger: *The Guardian* (12.11.94).
Commodification: see the excellent if depressing article by David Denby, 'Buried Alive,' *The New Yorker* (15.7.96), 48–58. See also Julian Stallabrass, *Gargantua: Manufactured Mass Culture* (London: Verso, 1996).
Tolkien on Disney: *Letters*, 17. The problem is not animation *per se*, by the way; none of these charges could seriously be laid against, say, Warner Brothers Cartoons, or Jim Henson's productions, let alone the literally marvellous work of Hayao Miyazaki.
On Walt Disney: see Marc Eliot, *Walt Disney, Hollywood's Dark Prince* (Birch Lane Press, 1993).
Carter and Tolkien: an aside: according to Susannah Clapp, Carter's literary executor, she gave her young son *The Hobbit* to read, but was less happy about *LoR*.
Carter's feminism: I am aware that her particular kind was unacceptable to puritanical and P.C. feminists, especially in North America.

136.
Northern air: Alison Lurie, 'Winter's Tales,' *The N.Y. Times Book Review* (19.5.96). Tolkien on drama: see 'On Fairy-Stories,' 46–8, 70–1.
Rushdie: quoted in *The New York Times Book Review* (8.3.92).

137.
Frank L. Baum, *The Wizard of Oz* (Ware, Herts.: Wordsworth Editions Ltd., 1993 [1900?]), 122–23. Since Tolkien's critics are rarely averse to *ad hominen* comparisons, let me forestall another one here by adding that Baum was a genocidal racist who advocated exterminating the last remaining Indians; see *Twin Light Trail: American Indian News* 2 (1992), 15.
Examples of mythic fiction: Charles Moorman, 'The Shire, Mordor, and Minas Tirith,' 201–17 in Isaacs and Zimbardo, *Critics*, 201, argues that *The Lord of the Rings* 'defines its own genre, just as *Moby Dick* does . . .'

138.
Unclassifiable: that is presumably what Humphrey Carpenter meant when he

said that Tolkien 'doesn't really belong to literature.' Carpenter's remarks on this occasion were disgracefully cavalier. ('Bookshelf,' broadcast on BBC Radio 4 on 22.11.91).

True company: this is of course simply a personal selection of the best mythic fiction (in my opinion — and again, I have obviously not read everything). Possibly Alan Garner's *Strandloper* (1996) should be added; at the moment, it's too soon to say.

Gold: *LoR*, I, 230.

## 6. Conclusion: Hope without Guarantees

139.
Epigraph: Kane, *Wisdom*, 256. *Cf.* a poet quoted by Enright, *Interplay*, p. 196: 'The poetry of earth is never dead.'
Three faces: 'Fairy-Stories,' 28.
Everlasting sea: Ehrenfeld, *Arrogance*, 269; and see Earle, *Sea Change*.
Escape: *LoR*, III, 224.

140.
'Positivist' etc.: Dowie, 'Gospel,' 283.
'Sanity and sanctity': *Letters*, 413.

141.
Hobbits' appreciation: *LoR*, III, 91.
Bath song: *LoR*, I, 142.
Essentials: *LoR*, III, 202; II, 402; III, 233, 259, 258.

142.
Richard Jefferies, *The Story of my Heart* (London: Longmans Green & Co., 1883), 4; as quoted in Mabey *et al.*, *Second Nature*, 119; and Jefferies, *Story*, 18, 98–9.
N. Frye: quoted by Camille Paglia in her *Sexual Personae* (London: Penguin, 1992), 223.
Scientific magician: 'Fairy-Stories,' 15.

143.
Tim Robinson, 'Listening to the Landscape,' *Irish Review* 14 (Autumn 1993), 21–32: 32; now reprinted in *Setting Foot on the Shores of Connemara and Other Writings* (Lilliput, 1996), 151–64. (*NB*: he is not responsible for my use of his term, with which he would ultimately disagree.)
Stories/narrative: see David Carr's important book *Time, Narrative, and History* (Bloomington: Indiana University Press, 1986).

144.
Kane, *Wisdom*, 50; my italics.
Gary Snyder, in *The Utne Reader* (5.6.95), 69. (And see, on the mythological

imagination and place, Roger O'Toole, 'Myth, Magic and Religion in Secular Literature: the Canadian Case,' *Journal of Contemporary Religion* 10:3 (1995), 297–307.)
Slavenka Drakulic, *The Observer* (21.11.93) and *The New Republic* (13.12.93).

145.
Sue Clifford and Angela King (Eds.), 7–18 in 'Losing Your Place,' in *Local Distinctiveness*, 11, 18, 7, 12. See also Sue Clifford and Angela King (Eds.), *From place to PLACE* (London: Common Ground, 1996).
'Organism': Dowie, 'Gospel,' 268 (quoting Mircea Eliade).

146.
'Kernel': *Letters*, 420, and see 221, 345; also Shippey, *Road*, 216. Tolkien was not consistent about this; e.g. 'Finnish . . . was the original germ of the Silmarillion': *Letters*, 87.
Imagined wonder: 'Fairy-Stories,' 18.
C. N. Manlove, *The Impulse of Fantasy Literature* (London: Macmillan, 1983), 156; cf. Attebery, *Strategies*, 16.

147.
Inside a song: *LoR*, I, 455.

148.
Political enchantment/magic: thus Ché Guevara ends up as a clothes boutique (along with Kalashnikov), the no-less-revolutionary Culpeper selling expensive herbal products, and so on.
Power of the Ring: *LoR*, III, 268.
Dominion of Men: *LoR*, III, 302 (and see 453).

149.
Recovery: 'Fairy-Stories,' 53.
The Shadow: *LoR*, I, 78.
'Estel': *LoR*, III, 423. See Tolkien's sad poem, 'The Sea-Bell.'
'Hope': *Letters*, 237.
Despair: the chief occupational hazard of people involved in defending Middle-Earth, probably followed closely by misanthropy.

150.
'Arda': *Morgoth's Ring*, 405.
A fairy tale: *LoR*, I, 257.

155.
Tolkien: *Letters*, 111, 116, 64.
Derrick Jensen and George Draffan, *Strangely Like a War: The Global Assault on Forests* (White River VT: Chelsea Green Publishing, 2003).

Verlyn Flieger, 'Taking the Part of Trees: Eco-Conflict in Middle-earth,' in *J.R.R. Tolkien and His Literary Resonance*, George Clark and Daniel Timmons (eds.), 147–58 (Westport CT: Greenwood Press, 2000). (With thanks to Dan Timmons for having the courtesy to send me a copy. And I speak for Jonathan Curry [p. 158], too.)

156.
David Wiggins, 'Nature, Respect for Nature, and the Human Scale of Values,' *Proceedings of the Aristotelian Society* XCX: 1–32, 10.
Tolkien on drama: 'On Fairy-Stories', 48.

159.
John Garth, *Tolkien and the Great War: The Threshold of Middle-earth* (London: HarperCollins, 2003).

# Bibliography

This lists the editions of Tolkien's works I have used plus the secondary sources I have found most helpful; it is not meant to be exhaustive. (I would also like to commend *The Guardian*'s excellent environmental reportage and analysis.)

Max Horkheimer and Theodor W. Adorno, *The Dialectic of Enlightenment* (New York: Continuum, 1994 [1944]).

Brian Attebery, *Strategies of Fantasy* (Bloomington, Indiana: University of Indiana Press, 1992).

Zygmunt Bauman, *Intimations of Postmodernity* (Oxford: Basil Blackwell, 1992).

Walter Benjamin, *Illuminations: Essays and Reflections*, ed. Hannah Arendt (New York: Schocken Books, 1969), 'The Storyteller,' 83–109.

Humphrey Carpenter (ed.), *Letters of J.R.R. Tolkien* (London: George Allen & Unwin, 1981).

David Ehrenfeld, *The Arrogance of Humanism* (Oxford: Oxford University Press, 1978).

Paul Ekins, *A New World Order: Grassroots Movements for Global Change* (London: Routledge, 1992).

Don D. Elgin, *The Comedy of the Fantastic: Ecological Perspectives on the Fantasy Novel* (Westport: Greenwood Press, 1985).

Kath Filmer, *Scepticism and Hope in Twentieth Century Fantasy Literature* (Bowling Green: Bowling Green State University Popular Press, 1992).

John Fowles, *The Tree* (St Albans: The Sumach Press, 1992; first published London: Aurum Press, 1979).

Fraser Harrison, 'England, Home and Beauty,' 162–72 in Richard Mabey with Susan Clifford and Angela King (eds.), *Second Nature* (London: Jonathan Cape, 1984).

Sean Kane, *Wisdom of the Mythtellers* (Peterborough, Ontario: Broadview Press, 1994).

Clyde Kilby, *Tolkien and The Silmarillion* (Berkhamsted: Lion Publishing, 1977).

Alkis Kontos, 'The World Disenchanted, and the Return of Gods and Demons,' 223–47 in Asher Horowitz and Terry Maley (eds.), *The Barbarism of Reason: Max Weber and the Twilight of Enlightenment* (Toronto: University of Toronto Press, 1994).

Ursula K. Le Guin, *The Language of the Night: Essays on Fantasy and Science Fiction*, ed. Susan Wood (London: The Women's Press, 1989).

Virginia Luling, 'An Anthropologist in Middle-earth,' in Reynolds and GoodKnight, *Proceedings*, 53–57 (see below).

Patricia Reynolds and Glen H. GoodKnight (eds.), *Proceedings of the J.R.R. Tolkien Centenary Conference* (Milton Keynes: The Tolkien Society, and Altadena: The Mythopoeic Press, 1995).

Mary Salu and Robert T. Farrell (eds.), *J.R.R. Tolkien, Scholar and Storyteller* (Ithaca NY: Cornell University Press, 1979).

T. A. Shippey, *The Road to Middle-Earth* (London: HarperCollins, 1992 [1982]).

J.R.R. Tolkien, *The Lord of the Rings* (London: Grafton Books, 1991).

———, 'On Fairy-Stories,' 9–73 in his *Tree and Leaf* (London: Unwin Hyman, 1988).

———, *The Hobbit* (London: Grafton Books, 1991).

———, 'Beowulf: The Monsters and the Critics,' in *The Beowulf Poet: A Collection of Critical Essays*, ed. Donald K. Fry, 8–56 (Englewood Cliffs: Prentice-Hall, 1968). Originally *Proceedings of the British Academy XXII*, 1936, 245–95).

Meredith Veldman, *Fantasy, the Bomb, and the Greening of Britain: Romantic Protest, 1945–1980* (Cambridge: Cambridge University Press, 1994).

# Index